ENVIRONMENTAL SCIENCE, ENGINEERING AND TECHNOLOGY

AQUIFERS: TYPES, IMPACTS AND CONSERVATION

ENVIRONMENTAL SCIENCE, ENGINEERING AND TECHNOLOGY

Additional books in this series can be found on Nova's website under the Series tab.

Additional E-books in this series can be found on Nova's website under the E-book tab.

WATER RESOURCE PLANNING, DEVELOPMENT AND MANAGEMENT

Additional books in this series can be found on Nova's website under the Series tab.

Additional E-books in this series can be found on Nova's website under the E-book tab.

ENVIRONMENTAL SCIENCE, ENGINEERING AND TECHNOLOGY

AQUIFERS: TYPES, IMPACTS AND CONSERVATION

ZOUBEIR OUAKILI AND HABIB CHIPPO
EDITORS

Nova Science Publishers, Inc.
New York

For permission to use material from this book please contact us:
Telephone 631-231-7269; Fax 631-231-8175
Web Site: http://www.novapublishers.com

LIBRARY OF CONGRESS CATALOGING-IN-PUBLICATION DATA

Aquifers: types, impacts, and conservation / editors, Zoubeir Ouakili and Habib Chippo.
p. cm.
Includes index.
ISBN 978-1-61942-091-5 (hardcover)
1. Aquifers. 2. Aquifers--Research. 3. Groundwater--Pollution. I. Ouakili, Zoubeir. II. Chippo, Habib.
GB1199.A69 2011
551.49--dc23
2011042517

Published by Nova Science Publishers, Inc. † New York

CONTENTS

PREFACE

Aquifers are typically saturated regions of the subsurface that produce an economically feasible quantity of water to a well or spring (e.g., sand and gravel or fractured bedrock often make good aquifer materials). Most land areas on Earth have some form of aquifer underlying them, sometimes at significant depths. In this book, the authors present current research in the study of the types, impacts and conservation of aquifers. Topics discussed include the effect of aquifer heterogeneity; hydrochemical features of groundwater from aquifer systems occurring in Sao Paulo, Brazil; aquifer system characterization using integrated geophysical methods; pollution risk of groundwater in a semi-arid region by wastewater rejections; a numerical study of aquifer thermal energy storage systems influenced by regional groundwater flow and fluid flow and contaminant propagation in fractured rock aquifers.

Chapter 1 - This study concerns the support for pump-and-treat remediation decision making. Proper input regarding the remediation process, its costs and outcomes is a prerequisite to the development of decision making tools, which should ultimately integrate the remediation process knowledge with real-time information about the domain of discourse and its goals. In this paper we study the effect of the formation heterogeneity on the remediation process of an aquifer whose top layers have been contaminated by light non-aqueous-phase liquid (LNAPL), namely fuels, like gasoline, kerosene, diesel fuel. We consider the parameters affecting treatment decisions, which include environmental, societal, health, economic parameters, and most importantly the characteristic of the remediation process. In order to identify characteristics of the pump-and-treat remediation of the heterogeneous formation like fractured permeable formation (e.g, fractured sandstone) we have considered a simplified model of a formation comprising of permeable blocks in which fractures are embedded. The formation is polluted by entrapped NAPL and the formation heterogeneity is examined for whether it can serve as an input parameter to the decision process. During the pump-and-treat process, NAPL is washed out quickly from the fractures because of their small storage. Therefore, most efforts of the remediation should be invested in cleaning-up the contaminated permeable blocks. Our theoretical analysis has indicated that characteristics of the remediation are determined by the preferential flow in the heterogeneous formation, which is quantified by a dimensionless number called "the mobility number" that represents the ratio of the permeable blocks' flow rate to that of the fracture network; and another parameter called "the dimensionless interphase mass transfer coefficient". Hence, these two parameters should be considered as input to the decision making process of the cost-effectiveness of remediation. Our theoretical modeling approach has quantitatively indicated that the heterogeneity of the permeable formation reduces the efficiency of the pump-and-treat remediation procedure, because the fracture network (parts of high permeability of the formation) allows bypassing of parts of the formation with low permeability (permeable blocks), which are contaminated by entrapped NAPL. In order to verify the theoretical modeling results we have constructed an experimental setup consisting of two parallel identical columns of rectangular cross sections, one was filled with homogeneous porous medium and the other was filled with the identical porous medium in which a fracture network was constructed. The columns were subject in parallel to simulated pump-and-treat remediation. The experimental results were in agreement with the theoretical modeling predictions.

Chapter 2 - Brazil owns the highest availability of renewable hydrological resources in world. However, in the beginning of the 21st century, it was the 26th country in terms of social water availability as a consequence of the great unequal distribution of the hydrological resources in the country. This is because it is mainly concentrated in northern and portions of northeastern and central western regions, where the population density is low. This advanced degree of unequal water distribution has caused special attention to the groundwater resources due to problems related to the interaction between the society and the environment. São Paulo is the most populous Brazilian state, comprising ~ 40 million inhabitants distributed over 645 municipalities. It has the highest number of industries and economic production, reaching 31% of the Brazilian GDP-gross domestic product. Despite the vigorous industrial production that includes high technology goods, the state also is well developed in agriculture and cattle breeding. This advanced stage of agricultural and industrial growth causes to São Paulo State an accentuated use of groundwater resources in water-supply systems. Several lithologies of the Paraná sedimentary basin outcrop in São Paulo State. They consist on sandstones, conglomerates, diamictites, tillites, siltstones, mudstones, limestones, shales, silex, rythmites, basalts, diabases and different types of Cenozoic covers like the recent deposits, terrace sediments and the Rio Claro Formation (sandstones, conglomerate sandstones and muddy sandstones). Many of these rock types are important reservoirs for groundwater in São Paulo State. Rio Claro city is located at the northeastern edge of the Paraná basin, cropping-out several units of the basin in the region. Rio Claro city is the most important municipality in Corumbataí River basin that extends over an area of about 1581 km^2. The area considered in this chapter includes Rio Claro city and portions of Araras, Leme, Cordeirópolis, Santa Gertrudes, Itirapina, Corumbataí and Ipeúna municipalities. Different groundwater systems have been focused in this chapter, among them: a phreatic aquifer comprising a cover of unconsolidated materials from Rio Claro Formation; sediments from Tubarão Group, comprising a confined aquifer about 40 m thick; deep bodies of diabase that store water within their fractures. In general, sedimentary aquifers exhibit different flow and hydrochemical characteristics in relation to that of the fractured rock aquifers and these aspects will be considered in this chapter.

Chapter 3 - The assessment of the hydrogeological conditions of a multilayered aquifer system requires a good knowledge of the aquifer geometry. The aquifer characterization is essential for the management of groundwater resources. This study has been carried out in the Mornag plain in North East of Tunisia and focuses on the geophysical methods application. It is a favourable site for assessing the performance and suitability of the geophysical methods for aquifer system identification and groundwater exploration. Thus, two complementary geophysical investigations comprising electrical resistivity and geophysical log analysis were carried out. This study provides a complete electrical image of the underground. Results show that the prospected area is characterized by the succession of several levels with contrasted resistivities, which are often affected by faults. This study should be useful for choosing the best sites for reconnaissance borings or test wells that will precede the exploitation of the aquifer system, in the future.

Chapter 4 - Mancha Oriental Aquifer (MOA) is one of the most extensive carbonate aquifers largest (7,260 km^2) in Southern Europe. MOA is located in the southeast of Spain and is encompassed within the Jucar River Basin. Since the 80's, the exploitation of groundwater resource has become a key driver for the socioeconomic development of MOA. Groundwater resources have been used to supply urban and industrial uses, and irrigation agriculture which nowadays area currently exceeds 100,000 ha. Groundwater withdrawals runs at rate of more than 400 Mm^3/yr. The overexploitation has caused two major quantitative impacts, namely a steady drop in groundwater level and a reduction of MOA discharge to Jucar River. In these sense, MOA is presented as a shining example of the problems that the exploitation of groundwater for agricultural purposes is for the sustainable management of water resources and ecosystems associated surface.

Chapter 5 - The present chapter is focused on the Djeffara aquifer system located in the south eastern part of Tunisia. This aquifer system, shared between Tunisia and Libya, constitutes the main water resource in all the coastal plain. Its recharge, particularly in its northern part is maintained only thanks to large-scale groundwater inflow from the CI aquifer system which constitutes the largest confined aquifer in the northern Sahara sedimentary basin. Under arid climatic conditions and an increasing water demand for irrigation, Djeffara aquifer system, presents different vulnerabilities to anthropogenic activities. Intensive exploitation of the aquifer during last years induced declining water levels, drying up springs and groundwater quality deterioration. In order to evaluate the factors that significantly influence groundwater quality, the multivariate statistical techniques were performed on 90 well representatives of groundwater samples gathered from Djeffara shallow aquifer. The collected samples were analyzed for physical-chemical parameters and stable isotope contents (^{18}O, ^{2}H). Principal component analysis (PCA) results identified three major principal components (PCs) representing 71% of cumulative variance. The contribution degrees of variation in a data set are 47% for 1st PC, 13% for 2nd PC, and 11% for 3rd PC of data set. The first factor represents salinization process with seawater intrusion. The second factor shows flushing of gypsum minerals from sedimentary rocks and nitrate contamination by irrigation return flow. The third factor revealed the mixing effect with deep groundwater of Continental Intercalaire aquifer occurring in the northern part of the basin. These results suggest that the groundwater quality in the study area seems to be controlled by not only saltwater intrusion but by other effects such as irrigation return flows and mixing with old groundwater of the Continental Intercalaire aquifer.

Chapter 6 - Surfactant-enhanced aquifer remediation (SEAR) is an innovative technology for the removal of nonaqueous phase liquids (NAPLs) in the aquifer. Heterogeneity in aquifer permeability constitutes one of the greatest challenges to remediation process because removal of NAPLs can occur only at the contacted formation by injected surfactant solution. A three-dimensional, multiphase, multicomponent compositional simulator is employed to simulate the solubilization and mobilization of NAPL during the remediation in spatially-correlated heterogeneous fields. Aquifer heterogeneity was accounted for by considering the permeability as a spatial random variable and a geostatistical method was used to generate random distributions of the permeability. Spatial distributions of NAPL saturations in the aquifer and temporal changes of organic recovery, effluent concentrations of organics and surfactant, and pressure drop at the injection well were compared with those in homogeneous aquifer and related to heterogeneity. Variations in permeability fields have a pronounced effect on the organic recovery efficiency due to the long-term persistence of nonaqueous phase liquid in bypassed portion of aquifer and additional dispersion. Higher heterogeneity also results in tailing of effluent organic concentrations and significant loss in injectivity over the remediation life. For a small slug, surfactant-enhanced remediation had a relatively small improvement on the recovery of NAPL, especially in highly heterogeneous aquifers. Migration of the high concentration organic plumes to the other layer by crossflow is also found to have a significant influence on SEAR behavior.

Chapter 7 - Groundwater is exposed, more and more intensively, to deliberate discharges of polluting effluents, sewage or storm water runoff in urban areas. Near cities, the sources of groundwater contamination are numerous and are related to many urban activities. The urban water is a source of contamination of groundwater by their concentration in organic and inorganic constituents. Other sources can be added such as air pollution, rain, washing pavements, etc... However, their origins can be complex and include stormwater, wastewater leakage networks. Leachates of municipal waste, septic tank are considered as a source of contamination and carry pollutants that have major environmental impacts. On Oued El Kebir basin in Tébessa (fig1) increased industrial and agricultural activities, and the alternative by-products of production or post-consumer waste, making them vulnerable groundwater resources. The plan amendments leading to variations in groundwater level by excessive pumping in urban areas can also cause contamination from interconnected flows. Impact may be a hydrological since a major exploitation of the resource has an action on the hydrodynamic behaviour of the aquifer. In this perspective, we try to examine the state of the groundwater resources of Tebessa plain, the last cut by the Oued El Kebir, which drains the wastewater from the entire urban area of the city and which location can contaminate groundwater.

Chapter 8 - This paper presents numerical investigations and thermohydraulic evaluation of two-well models of aquifer thermal energy storage (ATES) system operating under continuous flow regime. A three-dimensional numerical model for groundwater flow and heat transport is used to analyze the performance of thermal energy storage in the aquifer. Thisstudy emphasizes the effects of regional groundwater flow on the performance of the ATES system under various operation scenarios consisting of different parameters including flow conditions of injecting and producing water (temperature and flow rate), regional groundwater velocity and properties (hydraulic gradient), and aquifer characteristics (heat loss and size). Performances of ATES were compared in terms of temperature of extracted water and temperature field in the aquifer. The calculated temperature at the producing well varies within a certain range through the year and fluctuating quarterly a year, reflecting the seasonaltemperature conditions of injecting water. The pressure gradient across the system, which determines the direction and velocity of regional groundwater flow, has a substantial influence on the aquifer thermal storage. Water flow opposite to the direction of regional groundwater flow leads to an effective ATES because of smaller loss and less fluctuation in extracted thermal energy. Injection/production rate and geometrical size of the aquifer used in the model also impact the predicted temperature distribution at each stage and the recovery water temperature. But heat loss with the adjacent confining strata hasa minimal effect on the performance of ATES systems. The hydrogeological-thermal simulation is shown to be an integral part in the prediction of performance for a process as complicated as ATES systems.

Chapter 9 - It is known that there is a certain unpredictability associated with fractured and karstified rock formations that limits the comprehension of flow and transport phenomena that take place within them. Yet the wide spreading of these rocks makes it extremely important to improve their understanding in order to protect them from natural and anthropic hazards. In fractured rock aquifers the pronounced heterogeneity of the medium due to the existence of discontinuities may create hydrodynamic conditions of difficult interpretation and therefore imply uncertainty in modeling and in planning remediation interventions. In such context the Darcy law can be adapted to flow throughout fractures by means of the well known cubic law, valid under the assumption of ideal fractures, represented by smooth and parallel plates; however its application to real fracture systems characterized by rough and variable apertures has many times proved to be oversimplified and unrealistic. The parallel plate model is no more valid at higher depth where, due to compressive stresses, the walls of the fractures are pressed together forming asperities that lead to reduction in fracture aperture and cause flow channeling. In addiction, the partial closure of fractures lead to other relevant phenomena in the vadose zone: unsaturated portions within the fractures will act as barriers to liquid flow along them in that the water will tend to flow from one matrix block to another across the partially saturated fractures. In the vadose zone, the hydrodynamic structure of a fractured rock aquifer may be totally reversed: high flux zones at near saturation become low flux zones in unsaturated conditions. At full saturation however, in fractured aquifers joints act as major conduits for water, dissolved matter and contaminants; in highly heterogeneous fracture networks, open fractures as well as bedding planes, karstic conduits or faults may give even place to preferential flow paths for ground water, contaminants in solution, and free product. The relatively large size of fracture openings permit fluid to reach high velocities and to move quickly away from a leak site, contaminating large areas in a short time. Furthermore, as contamination moves through a fractured rock aquifer, it tends to diffuse from the flowing fracture water into the rock's essentially stagnant pore water. The rock matrix itself is often able to store ground water, contaminants in solution, or free product. These processes tend both to retard the plume's advance through a fractured rock mass and to substantially increase the difficulty of purging contamination from the aquifer: the cleanup of fractured rock aquifers in some cases requires many decades, even centuries. The parameters which most strongly govern the degree to which matrix diffusion prolongs the aquifer restoration process have been detected as rock's matrix porosity, fracture spacing, matrix diffusivity, the chemical identity of the contaminant(s), and the length of time the aquifer has been contaminated. It is straightforward that an ad-hoc cleanup technique has to act on the first three parameters that have been identified as crucial within the remediation process itself. Moreover, in fractured rock aquifers numerical models used to simulate remediation processes make use of averaged values for these parameters, as it is extremely difficult to determine them accurately. Making predictions of actual cleanup times is therefore most of the times of scarce significance. As planning effective remediation interventions requires spatial and temporal predictions on contaminant propagation, it proves to be necessary to implement specific interpretations of the classical methodologies of describing fluid flow and solute transport in fractured rock aquifers. The present chapter provides an analysis of the peculiar aspects that concern the dynamics of groundwater circulation and contaminant propagation in fractured rock aquifers for the application of the ad-hoc cleanup technique.

In: Aquifers: Types, Impacts and Conservation ISBN: 978-1-61942-091-5
Editors: Z. Ouakili, et.al, pp. 1-34

Chapter 1

SUPPORTING REMEDIATION DECISION MAKING: THE EFFECT OF THE AQUIFER HETEROGENEITY

Hillel Rubin[1,*], Sharon Yaniv[1], Eran Rubin[2,†], and Holger Schüttrumpf[3,‡]

[1]Faculty of Civil and Environmental Engineering, Technion – Israel Institute of Technology, Haifa 32000, Israel
[2]Department of Technology Management, Holon Institute of Technology, 52 Golomb Street, Holon 58837, Israel
[3]Institute of Hydraulic Engineering and Water Resources Management, RWTH Aachen University, D-52056 Aachen, Germany

ABSTRACT

This study concerns the support for pump-and-treat remediation decision making. Proper input regarding the remediation process, its costs and outcomes is a prerequisite to the development of decision making tools, which should ultimately integrate the remediation process knowledge with real-time information about the domain of discourse and its goals. In this paper we study the effect of the formation heterogeneity

* E-mail: hrubin@technion.ac.il.
† E-mail: eranrub@gmail.com.
‡ E-mail: schuettrumpf@iww.rwth-aachen.de.

on the remediation process of an aquifer whose top layers have been contaminated by light non-aqueous-phase liquid (LNAPL), namely fuels, like gasoline, kerosene, diesel fuel. We consider the parameters affecting treatment decisions, which include environmental, societal, health, economic parameters, and most importantly the characteristic of the remediation process. In order to identify characteristics of the pump-and-treat remediation of the heterogeneous formation like fractured permeable formation (e.g, fractured sandstone) we have considered a simplified model of a formation comprising of permeable blocks in which fractures are embedded. The formation is polluted by entrapped NAPL and the formation heterogeneity is examined for whether it can serve as an input parameter to the decision process. During the pump-and-treat process, NAPL is washed out quickly from the fractures because of their small storage. Therefore, most efforts of the remediation should be invested in cleaning-up the contaminated permeable blocks. Our theoretical analysis has indicated that characteristics of the remediation are determined by the preferential flow in the heterogeneous formation, which is quantified by a dimensionless number called "the mobility number" that represents the ratio of the permeable blocks' flow rate to that of the fracture network; and another parameter called "the dimensionless interphase mass transfer coefficient". Hence, these two parameters should be considered as input to the decision making process of the cost-effectiveness of remediation. Our theoretical modeling approach has quantitatively indicated that the heterogeneity of the permeable formation reduces the efficiency of the pump-and-treat remediation procedure, because the fracture network (parts of high permeability of the formation) allows bypassing of parts of the formation with low permeability (permeable blocks), which are contaminated by entrapped NAPL. In order to verify the theoretical modeling results we have constructed an experimental setup consisting of two parallel identical columns of rectangular cross sections, one was filled with homogeneous porous medium and the other was filled with the identical porous medium in which a fracture network was constructed. The columns were subject in parallel to simulated pump-and-treat remediation. The experimental results were in agreement with the theoretical modeling predictions.

Keywords: Heterogeneous porous medium, fractured formations, NAPL contamination, aquifer remediation, pump-and-treat remediation

INTRODUCTION

This study concerns supporting decision making regarding pump-and-treat remediation of an aquifer whose top layers have been contaminated by light non-aqueous-phase liquid (LNAPL), namely gasoline, kerosene, diesel fuel etc. Typical cases of aquifer contamination by LNAPL take place where the LNAPL is released close to the soil surface, infiltrates through the vadose zone, arrives at the water table and accumulates on top of the water table as LNAPL lenses. Later the LNAPL lenses spread horizontally; and due to seasonal and annual fluctuations of the water table the LNAPL penetrates into top layers of the free surface aquifer. In Israel, mainly the Coastal Plain aquifer (CPA) has been subject to such cases of contamination by different types of LNAPLs (e.g. Kanfi 1986, Rubin and Braester 2000, Rubin et al. 2008a). This is a free surface aquifer located underneath the most densely populated part of the country. The aquifer incorporates heterogeneous layers consisting of sand and fractured sandstone, namely heterogeneous formations.

Complicated decisions are involved in designing and implementing the remediation of the aquifer instead of leaving the aquifer subject to natural attenuation. Further, the decision about the proper method of the aquifer remediation should take into account the sustainability of this action. Namely, we postulate that the ultimate goal of remediation, i.e., effective sustaining of water resources, can be achieved by employing a decision process balancing among 4 domains: 1) Life & Public Health Domain, 2) Water Resources Domain, 3) Environmental Domain, and 4) Economic Domain. On top of it, under any option, if groundwater contaminated by dissolved and solubilized NAPL compounds is pumped out of the aquifer then the aquifer is subject to pump-and-treat remediation. As an example this phenomenon is a positive byproduct of operating hydraulic barriers aiming to control the spreading of contaminants in aquifers (e.g., Rubin et al. 2008b, 2009).

The heterogeneity the porous formation has often been considered the major component affecting the efficiency of the aquifer remediation. The objective of this chapter is to provide an overview of the parameters affecting treatment decisions by referring to pump-and-treat of a heterogeneous aquifer contaminated by entrapped NAPL. In order to identify characteristics of the pump-and-treat remediation of the heterogeneous formation like fractured permeable formation (e.g, fractured sandstone) we have considered a simplified model of a formation comprising of permeable blocks in which fractures are embedded. Due to fluctuations of the water table the formation is polluted by entrapped NAPL. During the pump-and-treat process, NAPL is washed out quickly from the fractures because of their small storage. Therefore, most efforts of the remediation should be invested in the cleanup of the contaminated permeable blocks. The theoretical modeling approach is later verified by carrying out a series of relevant experiments in the laboratory.

The Contaminated Heterogeneous Aquifer and Remediation Options

Figure 1 shows a schematic description of an aquifer made of a permeable fractured formation, which is contaminated by entrapped NAPL. According to this figure, NAPL is entrapped within all pores of the porous medium after a long time period. Due to the small storage of the fractures, the NAPL entrapped within the fracture network is washed out in a short time, whereas the NAPL entrapped within the permeable blocks of the formation persists as a chronic source polluting the flowing groundwater.

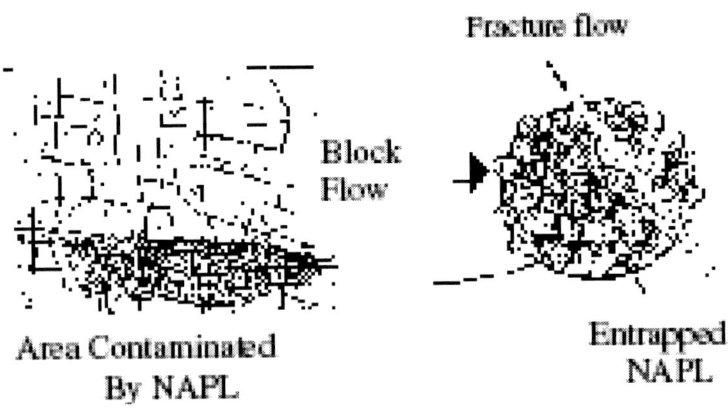

Figure 1. Schematic description of aquifer comprised of fractured formation contaminated by entrapped NAPL.

Contamination of the aquifer by entrapped LNAPL originates from seasonal and annual fluctuations of the water table. Such fluctuations may be attributed to rain, draught, and also intensive pumping. We refer to the case of the CPA significant contamination in 1991/2 due to a winter rich with rain storms that caused 6 m rise of the water table leading to entrapment of a kerosene lens 70 cm thick with horizontal size of 600 × 300 m^2 floating on top of the water table (Rubin and Braester 2000) in the top layers of the aquifer.

The event of the CPA water table rise in winter 1991/2 led to a kerosene (LNAPL) entrapment within the top 6 m layer of the aquifer, which consists of fractured sandstone.

Basically complicated decisions are involved in implementing remediation of the heterogeneous aquifer rather than allowing Mother Nature to carry out the natural attenuation (Wiedemeier et al. 1999). Further, the decision about the proper method of the soil and aquifer remediation should take into account the sustainability of this action. Such an approach incorporates the adequate balance among 4 domains:

1. Life and human health domain,
2. Environmental domain,
3. Water resources domain, and
4. Economic domain.

Therefore, the technology issue of applying sustainable remediation should be the outcome of a decision making process leading to that balance. In general, we postulate that remediation is definitely considered wherever there is an immediate risk to human health due to the presence of contaminants in the environment. However, when the aquifer remediation is considered for enriching or preserving regional water resources, in many cases applying a limited effort of capture by hydraulic barriers (e.g., Rubin et al. 2008b, 2009) is sufficient and complies with requirements of the economic domain. These two possible approaches illustrate that sustainable remediation is basically incorporated with management and decision making processes, as well as management of technologies. In essence remediation comes to repair the contaminated soil and aquifer. Pump-and-treat is a common procedure applied to provide proper remediation of the aquifer contaminated by various types of contaminants. This procedure is eventually also applied to keep operational hydraulic barriers (Rubin et al. 2008b, 2009). In this study we consider the effect of the aquifer heterogeneity on the efficiency of the pump-and-treat remediation procedure. We evaluate the parameters which can facilitate the remediation decision process and how they may interact with the four domains. The model representing the heterogeneous aquifer in this study is termed "fractured permeable formation". We introduce the basic conceptual model of such a heterogeneous aquifer, analyze its remediation characteristics by applying numerical simulations, and later carry out experiments that provide experimental results supporting the numerical simulations.

BASIC MODEL OF THE AQUIFER MADE OF PERMEABLE FRACTURED FORMATION

Figure 2 describes the basic model of the aquifer consisting of a permeable fractured formation that part of it is contaminated by entrapped NAPL. According to the previous section, this aquifer is subject to pump-and-treat remediation.

According to Figure 2, the formation is eventually a heterogeneous porous medium. Part of the groundwater flow is called "fracture flow" and is carried out through the fracture network, and another part, called "block flow" is carried out through the permeable blocks of the formation, in which the fractures are embedded. Figure 2 introduces a two dimensional (2D) model of the flow and transport domain. The storage of the fracture network is much smaller than that of the permeable blocks, and also the flow velocity through the fractures is much faster than through the blocks; therefore, we assume that the NAPL entrapped within the fractures is washed out very quickly. On the other hand NAPL entrapped within the permeable blocks is persistent as a chronic source of groundwater pollution for a long time period, and most efforts of the aquifer remediation should be invested in cleaning-up the blocks. The block clean-up is carried out by gradual dissolution and solubilization of the entrapped NAPL and organic mass transfer from the entrapped NAPL into the flowing groundwater.

Figure 2 refers to 3 stage conceptual 2D model:

Stage (a):	Macroscale - the aquifer that is partly contaminated by entrapped NAPL,
Stage (b):	Subdomain with thickness of a single block,
Stage (c):	Elementary fracture volume.

Our modeling development refers to the subdomain of Figure 2, through which flows the total discharge Q_t of groundwater in the longitudinal direction that is symbolized by the coordinate x^*. The total subdomain discharge of groundwater incorporates the fracture discharge Q_f and the block discharge Q_b. It should be noted that the fracture discharge is carried out in the direction of the fracture, whereas the block discharge is carried out in the direction of the hydraulic gradient of the aquifer flow, namely in the longitudinal x^*-direction.

In parts of the aquifer free from entrapped NAPL the fracture and block discharges are constant with values Q_{f0} and Q_{b0}, respectively. In parts of the aquifer contaminated with entrapped NAPL the relationships between Q_t, Q_f and Q_b are given by:

$$Q_t = Q_f + Q_b \quad , Q_b = \int_0^{B\sin\theta} q_b dy^* \tag{1}$$

where y^* is the vertical coordinate, and q_b is the specific discharge of the block flow that takes place mainly in the longitudinal x^*-direction.

The elementary fracture volume shown in Figure 2 helps to analyze the transport of dissolved organic mass from the block flow into the fracture flow and along the fracture network. In this figure C^* is the solute concentration in the fracture flow, is the solute concentration of the block flow entering the elementary fracture volume, x^* and y^* are the longitudinal and vertical coordinate, respectively. Assuming the flow in the subdomain is steady and neglecting second order terms, we obtain for the mass conservation of water and organic solute within the elementary fracture volume:

$$\frac{\partial Q_f}{\partial x'} dx' + \left(q_{bex} - q_{ben}\right) dy^* = 0 \tag{2}$$

$$\frac{\partial \left(Q_f C^*\right)}{\partial x'} dx' + \left(q_{bex} C^* - q_{ben} C_b^*\right) dy^* = 0 , \tag{3}$$

where q_{ben} is the block specific discharge entering the elementary fracture volume, q_{bex} is the block specific discharge exiting the elementary fracture volume.

The connections between the coordinates of the subdomain and those of the elementary fracture volume lead to the following relationships:

$$dx^* = dx' \cos\theta; \quad dy^* = dx' \sin\theta . \tag{4}$$

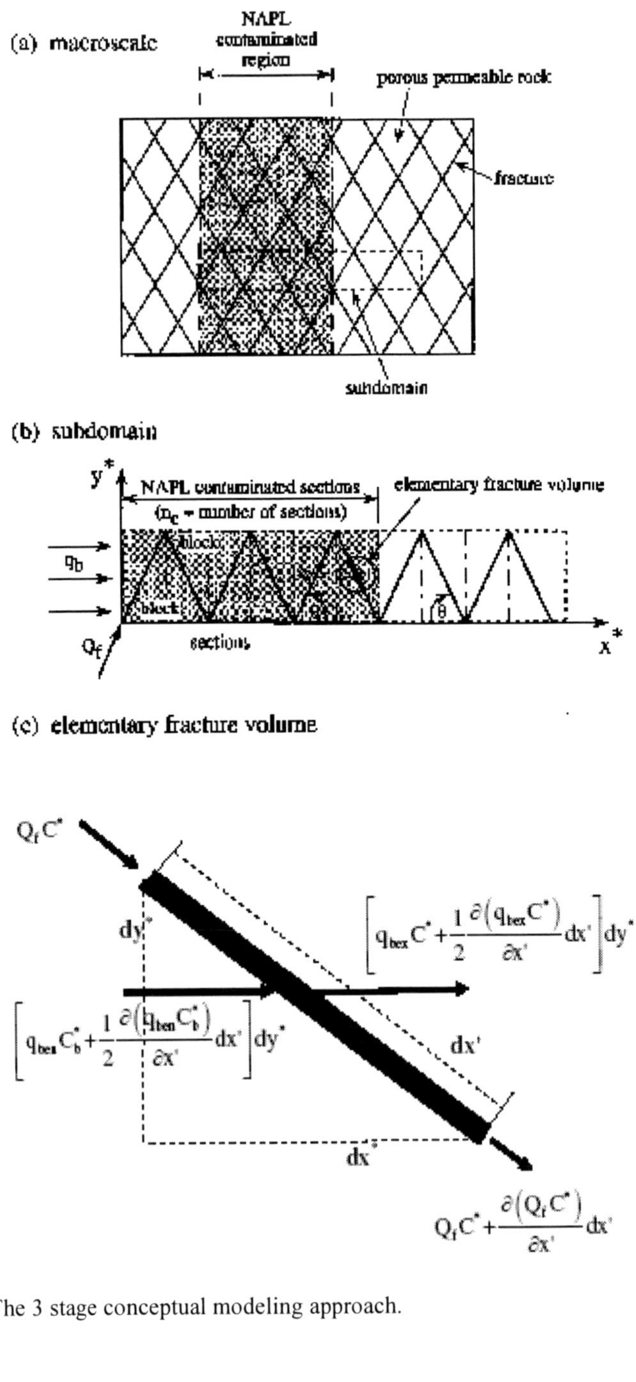

Figure 2. The 3 stage conceptual modeling approach.

We multiply Eq. (2) by C^* and introduce the product into Eq. (3) and apply the relationship of Eq. (4) to obtain:

$$Q_f \frac{\partial C^*}{\partial x'} dx' + q_{ben} \left(C^* - C_b^* \right) dx' \sin\theta = 0 . \quad (5)$$

This expression leads to:

$$Q_f \frac{\partial C^*}{\partial x^*} + q_{ben} \tan\theta \left(C^* - C_b^* \right) = 0 . \quad (6)$$

Eq. (6) is the differential equation describing changes of the organic solute along the fracture network of the subdomain shown in Figure 2. Such changes are attributed to organic mass transfer between the fracture and the block flows. It should be noted that due to the assumed small storage of the fracture network Eq. (6) does not incorporate any time dependent term.

Eq. (6) is simplified by applying the following dimensionless coordinates and domain variables:

$$x = \frac{x^*}{B\cos\theta}; \; y = \frac{y^*}{B\sin\theta}; \; t = \frac{t^* V_{b0}}{B\cos\theta}; \; C = \frac{C^*}{C_s^*}; \; C_b = \frac{C_b^*}{C_s^*}, \quad (7)$$

where is the solute equilibrium concentration in the water phase, V_{b0} is the average interstitial flow velocity of the water phase in blocks free from entrapped NAPL, t^* is time, which will be applied to carry out calculations concerning changes in time within the aquifer. Dimensional coordinates and variables are represented by symbols with upper asterisks. Dimensionless variables are represented by the same symbols without asterisks.

By introducing the dimensionless coordinates and variables of Eq. (7) into Eq. (6) we obtain the following dimensionless differential equation of solute concentration changes along the fracture network of the subdomain shown in Figure 2:

$$\frac{\partial C}{\partial x} + N_M C = N_M C_b , \quad (8)$$

where N_M is the mobility number, which is defined by:

$$N_M = \frac{q_B B \sin\theta}{Q_f}, \tag{9}$$

where Q_f is the local discharge of the fracture flow, q_B is the local value of the specific discharge of the block flow that enters the fracture where the value of N_M is calculated. In parts of the aquifer contaminated by NAPL values of Q_f and q_B are subject to changes along the fracture and also in time; because during the remediation process organic mass is subject to gradual transfer from the entrapped NAPL into the block flow and from the block flow into the fracture flow. Therefore, in parts of the aquifer contaminated by entrapped NAPL, the mobility number is subject to changes in time and along the domain. On the other hand in parts of the domain free from entrapped NAPL, the mobility number is constant, and it represents the ratio of the block flow discharge to the fracture flow discharge, namely:

$$N_{M0} = \frac{q_{b0} B \sin\theta}{Q_{f0}}. \tag{10}$$

where Q_{f0} is the fracture flow discharge in parts of the aquifer free from entrapped NAPL. The symbol Q_{b0} refers to the block flow in parts of the aquifer free from entrapped NAPL. We may apply the value of N_{M0} as the parameter that represents the preferential flow in the domain. According to ranges of the mobility number we may classify the following types of formations (Birkhölzer et al. 1993a):

1. Homogeneous formations $N_{M0} > 25$
2. Fractured permeable formations 0.01 N_{M0} 25
3. Fractured porous formation $N_{M0} < 0.01$

In homogeneous formations, fractures are rare and effects of preferential flow are negligible. In fractured porous formation groundwater flow is dominated by the fracture flow and the flow through the porous blocks is very small. The present study mainly concerns groundwater flow attributed to fracture and block flows of similar orders of magnitude, namely cases of fractured permeable formations. The mobility number is one of the parameters that should be considered when analyzing and designing the remediation of a heterogeneous porous medium by the pump-and-treat procedure. The value of this parameter provides a possible measure of the effect of the heterogeneity caused by the presence of fractures on the aquifer remediation.

Except for the effect of the preferential flow on the remediation process, in order to analyze the effect of the formation heterogeneity on the pump-and-treat procedure we should refer to parameters of transfer of organic mass from the entrapped NAPL into the flowing groundwater. In the permeable blocks there are two fluid phases: the flowing water phase and the entrapped NAPL phase. Therefore, we may introduce:

$$S_w + S_n = 1, \tag{11}$$

where S_w is the water saturation, and S_n is the saturation of the NAPL entrapped within the permeable blocks.

Solute dispersion during the flow through the fractured permeable formation is mainly attributed to mixing between the fracture and the permeable block flows (Birkhölzer et al. 1993a, b). Therefore, by applying the mass conservation for the water phase and the organic solute in the permeable block flow, respectively we obtain:

$$\frac{\partial(\phi_b S_w)}{\partial t^*} + \frac{\partial q_{bx}}{\partial x^*} + \frac{\partial q_{by}}{\partial y^*} = 0 \tag{12}$$

$$\frac{\partial\left(\phi_b S_w C_b^*\right)}{\partial t^*} + \frac{\partial\left(q_{bx} C_b^*\right)}{\partial x^*} + \frac{\partial\left(q_{by} C_b^*\right)}{\partial y^*} = k_f\left(C_s^* - C_b^*\right). \tag{13}$$

where q_{bx} is the longitudinal component of the permeable block specific discharge in the longitudinal x^*-direction, q_{by} is the vertical component of the permeable block specific discharge in the vertical y^*-direction, and k_f is the lump coefficient of mass transfer from the entrapped NAPL into the water phase.

It is common to apply the Sherwood number for characterizing the effect of the domain and the NAPL characteristics on the value of k_f. The Sherwood number is defined as:

$$\mathrm{Sh} = \frac{k_f d^2}{D} \quad ,(14)$$

where d is the characteristic particle size of the porous medium, D is the diffusivity of the fluid. By using data of many theoretical and experimental studies Rubin et al. (2008a) showed that the Sherwood number can approximately be represented by:

$$\mathrm{Sh} = \beta_0 S_n^{\beta_1} . \tag{15}$$

where β_0 is a dimensionless parameter depending on the characteristic particle size of the porous medium, and β_1 is a dimensionless power coefficient whose value varies between 0.7 and 1.1.

We subtract Eq. (12) from Eq. (13) while neglecting the vertical component of the block specific discharge, namely assuming $q_b \approx q_{bx}$, and representing the result in dimensionless format; in this manner we obtain:

$$\frac{\partial C_b}{\partial t} + v_r \frac{\partial C_b}{\partial x} = K_f \left(1 - C_b\right), \tag{16}$$

where v_r is the relative flow velocity, namely the ratio of average interstitial flow velocity in pores contaminated by entrapped NAPL to that velocity in pores free from entrapped NAPL, and K_f is the dimensionless interphase mass transfer coefficient. The two dimensionless parameters described in the preceding phrase are defined by these expressions, respectively:

$$v_r = \frac{V_b}{V_{b0}}; \; K_f = \frac{k_f B \cos\theta}{q_{b0} S_w} . \tag{17}$$

where V_b is the average interstitial flow velocity in blocks contaminated by entrapped NAPL, and V_{b0} is the average interstitial flow velocity in blocks free from entrapped NAPL.

According to Eq. (16) the permeable block flow accumulates organic solute from the NAPL that is entrapped within the block's pores. When the block flow arrives at the fractures it mixes with the fracture flow, as described in Figure 2. According to Eq. (17), the dimensionless interphase mass transfer coefficient is the parameter governing the process of organic mass transfer from the entrapped NAPL into the flowing water phase. This parameter and the mobility number govern the remediation of the heterogeneous domain by the pump-and-treat procedure. According to Eqs. (15) and (17) during the remediation process the value of K_f is subject to changes according to the following expression:

$$K_f = K_{f0}\left(\frac{S_n}{S_{n0}}\right)^{\beta_1}\left(\frac{S_{w0}}{S_w}\right) = K_{f0}\left(\frac{S_n}{S_{n0}}\right)^{\beta_1}\left(\frac{1-S_{n0}}{1-S_n}\right), \tag{18}$$

where lower index zero represents the beginning of the remediation process.

By considering the conservation of the organic mass (NAPL and organic solute) in an elementary volume of the permeable block we obtain:

$$\phi_b \rho_n \frac{\partial S_n}{\partial t^*} + q_b \frac{\partial C_b^*}{\partial x^*} = 0, \tag{19}$$

where ρ_n is the density of the NAPL.

By introducing the dimensionless variables of Eq. (7) into Eq. (17) we obtain:

$$\frac{\partial S_n}{\partial t} + q_r C_{nv} \frac{\partial C_b}{\partial x} = 0. \tag{20}$$

where q_r is the ratio of q_b to q_{b0}, and C_{nv} is the volumetric solute concentration of equilibrium, which is defined by:

$$C_{nv} = \frac{C_s^*}{\rho_n}. \tag{21}$$

We refer to the remediation process in an aquifer represented by the subdomain shown in Figure 2. For this subdomain the average entrapped NAPL saturation in every cross section and the flux average solute concentration (the average flux concentration incorporates the solute flux in the fracture and the block flows) are respectively defined by:

$$S_{nav} = \frac{1}{B\sin\theta}\int_0^{B\sin\theta} S_n dy^* \tag{22}$$

$$C_{av}^* = \frac{Q_f C^* + \int_0^{B\sin\theta} q_b C_b^* dy^*}{Q_t} = \frac{Q_f C C_s^* + C_s^* B \sin\theta \int_0^1 q_b C_b dy}{Q_t}. \quad (23)$$

By introducing the dimensionless variables of Eq. (7) into Eqs. (22) and (23), respectively we get:

$$S_{nav} = \int_0^1 S_n dy . \ (24)$$

$$C_{av} = \frac{1}{1 + N_{M0} \int_0^1 k_{rw} dy} \left[C + N_{M0} \int_0^1 C_b k_{rw} dy \right]. \quad (25)$$

where k_{rw} is the relative water permeability of the block, which depends on the degree of water saturation in the permeable block.

During the remediation of the subdomain shown in Figure 2, the value of the solute concentration in the water flux is subject to changes. By integrating Eq. (25) over time we get the description of the remediation process and which part of the subdomain contaminated by entrapped NAPL is subject to clean-up by the groundwater flowing through the subdomain during the pump-and-treat procedure.

Eqs. (10) and (18) define the two main parameters that govern the remediation of the particular type of heterogeneous porous medium defined as fractured permeable formation. These parameters are the mobility number and the dimensionless interphase mass transfer coefficient. The numerical simulations carried out by Rubin et al. (2008a) indicated that major effects on the clean-up of fractured permeable formation contaminated by entrapped NAPL are attributed to the dimensionless interphase mass transfer coefficient. The effect of the mobility number, which is connected with the preferential flow, is significantly smaller than that of the dimensionless interphase mass transfer coefficient. However, the numerical simulation predictions should be evaluated by experimental results.

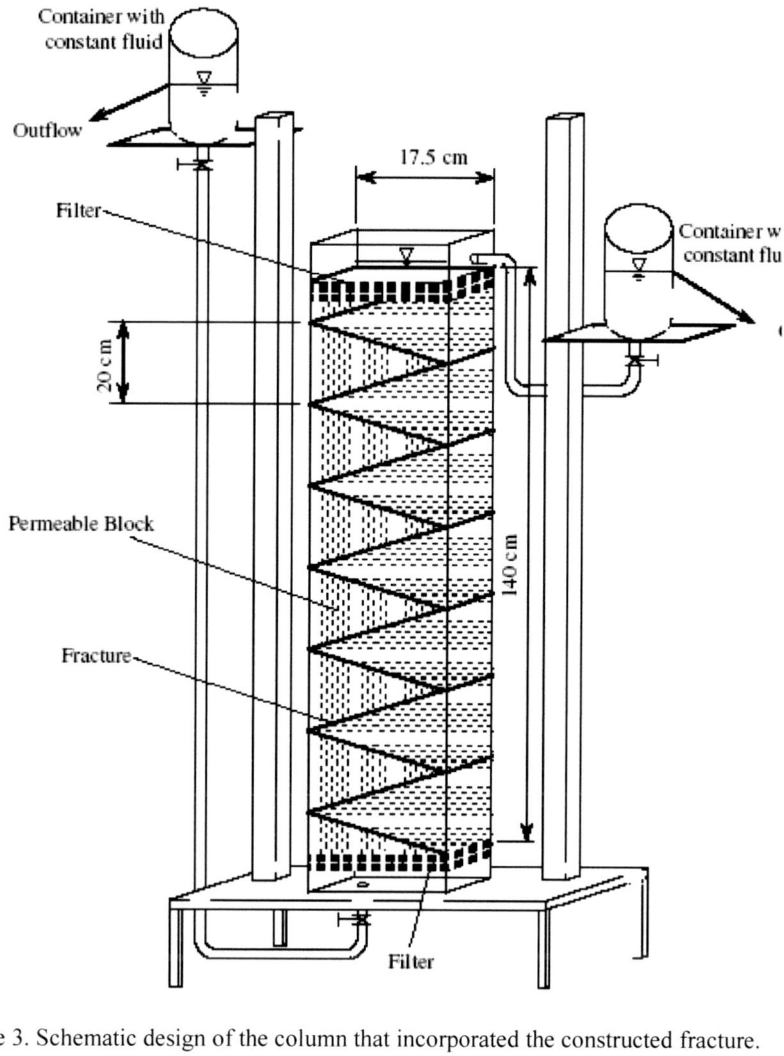

Figure 3. Schematic design of the column that incorporated the constructed fracture.

THE EXPERIMENTAL SET-UP

Figure 3 shows the schematic design of the experimental set-up used for carrying out the laboratory experiments. This set-up is based on using 2 identical columns of rectangular cross section made of glass 150 cm high, 17.5 cm wide, and 6 cm thick, filled with homogeneous porous medium. In one column a model of a continuous fracture similar to that of the subdomain shown in Figure 2 has been constructed. Therefore our experimental setup incorporated a homogeneous column and a heterogeneous (fractured) column. We have used two types of homogeneous porous medium to fill both columns:

1. Glass beads with characteristic particle $d = 2$ mm, and
2. Sand from the shore of Ashdod, Israel ("Ashdod Sand").

The Ashdod Sand properties were: porosity 0.375, and hydraulic water conductivity 0.226 cm/sec (around 20 m/day).

Figure 4. Construction of the porous column with the fracture model.

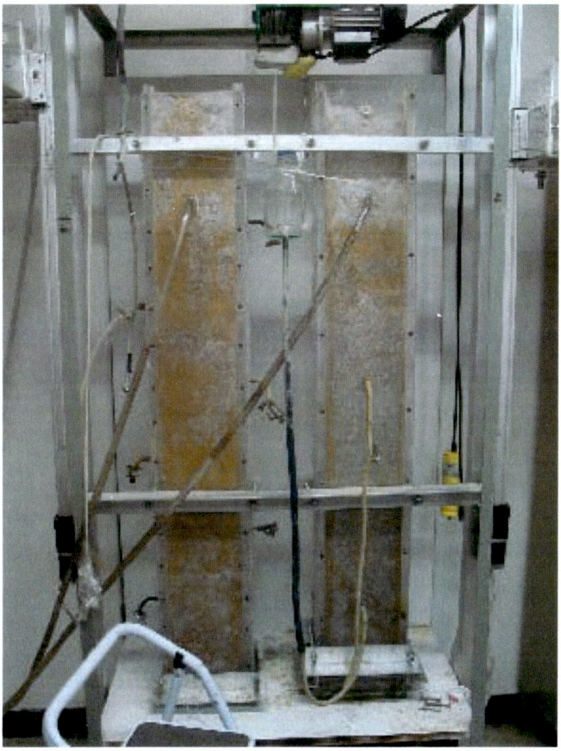

Figure 5. An experiment with the 2 parallel columns.

We have carried out many preliminary tests in order to determine the type of material that might fit for constructing the model of the fracture network in the column filled with homogeneous porous medium to represent the fractured permeable formation. Finally, it was found convenient to construct the fracture model from smooth pebbles whose characteristic diameter was d = 2 cm, mixed with glass beads (d = 2 mm), which were laid in a narrow channel extended as a broken line along the column, as shown in Figure 3. Figure 4 describes the construction of the fracture model from pebbles mixed with glass beads in the column made of Ashdod Sand.

Parallel to the construction of the column that incorporated the constructed fracture model is an identical column of the same homogeneous porous material that has been constructed for the same set of experiments. Therefore, we could carry out experiments in parallel with two identical columns. One column was filled with homogeneous porous medium and the other one was filled with the same porous medium in which a model of fracture has been constructed. Both columns were subject to identical experimental conditions (input and output head boundary conditions), as described in the next section.

Figure 5 shows an example of 2 parallel columns of the experimental set-up made of Ashdod Sand (one column is homogeneous, the other one is heterogeneous, due to the constructed fracture) during the experiment.

CARRYING OUT THE EXPERIMENTS

Before carrying out the experiments all parts of the experimental set-up and systems were subject to preliminary testing procedures in order to verify their proper operation. Special arrangements were required to avoid any leakage of kerosene, because experiments started with flow of water, kerosene and water with kerosene through the experimental set-up.

After evaluating the proper operation of the experimental set-up and all parts of the experimental systems, we determined the ranges of dimensionless parameters under which we could operate the experimental set-up and carry out the experiments. We found that experiments with a porous medium made of glass beads could be carried out in the range of Reynolds numbers between 0.5 to 1.5, namely at the upper limit of the linear Darcy law. Therefore, in the column with constructed fractures made of only pebbles with particle size 2 cm the flow was outside the linear range of Darcy law. According to Eq. (10) in order to change the mobility number of the heterogeneous column of a given type of permeable block material (e.g., Ashdod Sand) it is possible to change the structure of the constructed fracture, namely, the width of the constructed fracture, its orientation and different mixtures of pebbles with glass beads. By using such options, the experiments could be carried out in the mobility number range of 0.8 N_{M0} 2.

After completing the set of preliminary tests of flashing homogeneous and heterogeneous columns with water, we started the stage of entrapping NAPL within these columns and by flushing them with water for clean-up. As described in the following sections, two series of experiments have been carried out.

EXPERIMENTS OF SERIES A

In experiments of series A, NAPL (kerosene) has been entrapped by dropping and raising the water table. In this series of experiments, we filled both parallel columns (one column is homogeneous, and the other one is heterogeneous due to the constructed fracture) with water. The water was supplied very slowly to each column through its bottom valve in order to avoid entrapment of air within the pores of the blocks and the constructed fracture, until the columns had been completely filled with water. Then the water supply valve at the bottom of each column was shut off. At the top of each column we connected a container with a measured quantity of kerosene and started slowly draining water from the bottom of the columns while measuring the quantity of drained water and keeping constant level of kerosene in the container by supplying the kerosene and constantly measuring the quantity of kerosene added to the container. From time to time we took samples of the water drained from the bottom of the columns. When the experiment started pure water was drained. As the experiment progressed the samples indicated whether pure water was still draining out from the column or whether a mixture of water with kerosene was starting to drain out. When kerosene started to appear with the water drained from the column, the drainage from the column stopped. Then we calculated the quantity of kerosene located within the column, which was saturated with water and kerosene. At that time some parts of the kerosene were free and others were entrapped within pores of the porous medium. Therefore, after stopping the drainage from the bottom of the columns, the experimental set-up was kept with no experimental activity. We waited 3 days to allow redistribution and development of steady state of fluid phase distribution in the pores of the permeable blocks and the constructed fracture. After the 3 days, the 2 columns were reconnected to the source of water, and became subject to an upward constant hydraulic gradient, leading to water flow flushing the columns in an upward direction, while both columns were subject to identical hydraulic gradients. While flushing an upward flow of water through the columns, we collected samples of the fluid at the top outlet of each column. By determining quantities of kerosene in these samples we calculated the quantities of kerosene washed out from the columns. In the beginning only kerosene was washed out from the columns. When samples of kerosene mixed with water started to appear, we assumed that all quantities of free kerosene had been washed out, and the clean-up process of the entrapped kerosene started. Under such conditions we calculated the quantities of entrapped kerosene in the columns. In the heterogeneous columns, we assumed that all of the entrapped kerosene was located within the permeable blocks, because the constructed fracture was comprised of coarser material so the flow velocity was much faster.

After all of the free kerosene had been washed out from the columns, the experiments started to simulate the pump-and-treat remediation of the aquifer. Samples were taken every 1 to 3 hours for a time period of 3 days. Each sample was taken to a measuring tube of volume 100 mL into which a quantity of 20 gr NaCl was added before collecting the sample of water with kerosene. Then the sample was refrigerated for 12 hours to separate the kerosene from the water phase; the floating phase was the kerosene. The kerosene phase was taken out by Pasteur pipette, and the remaining fluid was weighted to determine the quantity of water in the sample. From all measurements carried out with the samples we calculated the quantities of kerosene washed out from each one of the columns. By carrying out the experiments with the parallel two columns, one column homogeneous and the other heterogeneous, we could compare differences and effects of the porous medium heterogeneity on the aquifer remediation by the pump-and-treat procedure.

EXPERIMENTS OF SERIES B

In experiments of series B, NAPL (kerosene) has slowly been supplied through the bottom of the two parallel columns (one column was homogeneous, and the other one was heterogeneous due to the constructed fracture) in order to avoid entrapment of air and lead to complete saturation of the pores with kerosene until the kerosene reached the top outlets of the columns. Then each column was disconnected from the source of kerosene, and the bottom was connected to a source of water, to start the water wash flow in the upward direction in both columns to clean-up the columns from the free kerosene. During the wash flow, we collected samples to assure free kerosene was completely washed out. When mixtures of kerosene with water started to appear we assumed that free kerosene had been washed out, and allowed 3 days for fluid phase redistribution and steady state entrapment of kerosene in the blocks. After 3 days, we started the experimental part of simulating the pump-and-treat procedure, as described in the preceding section.

Experimental Results with Columns Made of Glass Beads

Figure 6 describes the results of organic solute concentration changes in water at the top outlet of both columns made of glass beads subject to clean-up.

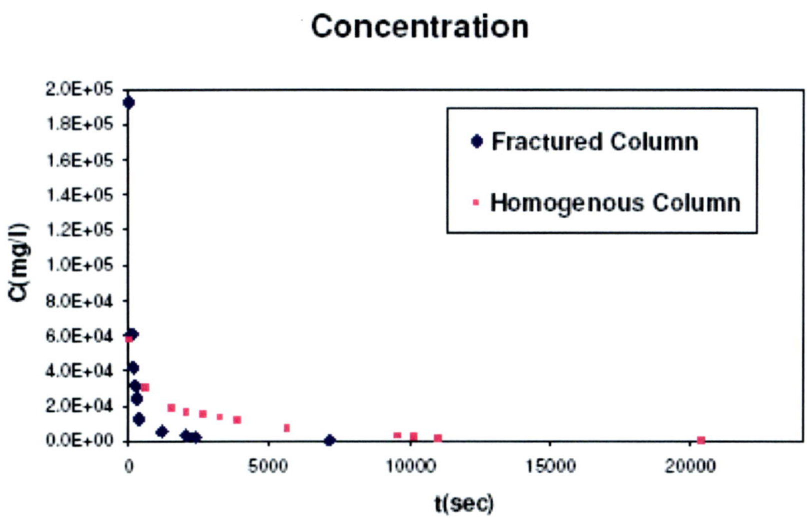

Figure 6. Changes of organic solute concentration in columns made of glass beads.

Figure 7 describes the results of entrapped kerosene saturation changes due to the clean-up of both columns made of glass beads.

It should be noted that solute concentrations shown in Figure 6, as well as entrapped kerosene saturations shown in Figure 7, were extremely low and they were continuously decreasing during the clean-up process. It means that quantities of entrapped kerosene in both columns made of glass beads were very minor. Both columns were subject to identical hydraulic gradients. However, the constructed fracture made of pebbles mixed with glass beads increased the hydraulic conductivity of the expected to be heterogeneous (fractured) column. Therefore, the total discharge flowing through the heterogeneous column was larger than that flowing through the homogeneous column. However, the mobility number in the heterogeneous column was 2, namely most of the flow was still carried out through the permeable blocks. The final result was that the solute concentration decreased faster in the heterogeneous (fractured) column than in the homogeneous one. It seems that because of the high Reynolds number of the flow in the glass bead medium and the constructed fracture made of pebbles and glass beads the effect of the medium heterogeneity was not significant in the set of experiments carried out with columns made of glass beads, and unexpectedly, as indicated by Figures 6 and 7, the clean-up of the heterogeneous (fractured) column made of glass beads was faster than that of the homogeneous column made of glass beads. Basically the clean-up characteristics in both columns had identical characteristics. Possibly, in this case the saturation of kerosene entrapped within the constructed fracture was of a value similar to that of the kerosene entrapped in the permeable blocks, and the constructed fracture did not introduce typical features of heterogeneity into the column made of glass beads.

We may conclude that the experimental results with columns made of glass beads could not be used for evaluating the effect of formation heterogeneity on the aquifer remediation, because quantities of entrapped NAPL were very minor, and possibly entrapment within the constructed fracture was of an order of magnitude similar to that within the permeable blocks. Maybe there are field cases in which observed heterogeneity of the formation does not necessarily lead to remediation characteristics typical of heterogeneous porous medium.

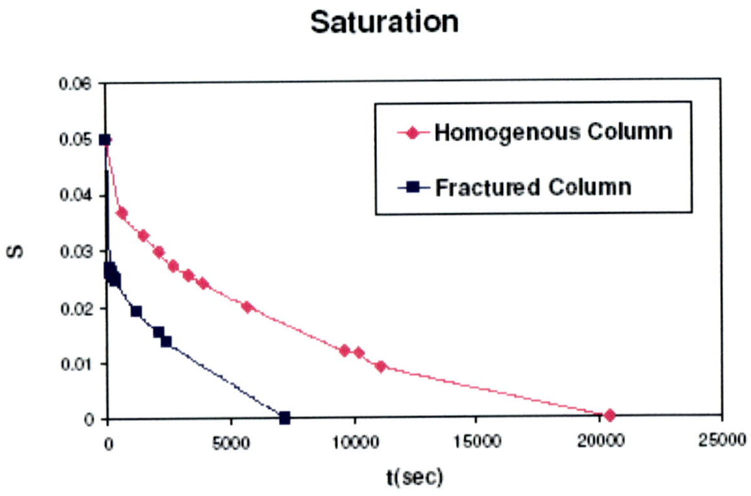

Figure 7. Changes of kerosene saturation in columns made of glass beads.

EXPERIMENTAL RESULTS WITH COLUMNS MADE OF ASHDOD SAND

The initial saturation of entrapped kerosene in columns made of Ashdod Sand was around 0.3. The mobility number in the heterogeneous column was around 0.8, namely more than half of the water flow was carried out through the constructed fracture.

Figure 8 describes changes in time of the organic solute at the outlet of the test section during the remediation of both columns made of Ashdod sand. In Figure 8 we applied the criterion of number of pore volumes for describing the rate of progress of the remediation instead of the dimensionless time that we used in Figures 6 and 7, as used in the studies of Geller and Hunt (1993) and also by Nambi and Powers (2000, 2003). In the next section, we find it appropriate to compare results of our study with those of Geller and Hunt (1993) and also by Nambi and Powers (2003).

According to Figure 8, in the homogeneous column, the outlet solute concentration was continuously decreasing during the clean-up process. Further, in the case of the homogeneous column the complete clean-up process was accomplished within a smaller number of pore volumes than that in the case of the heterogeneous (fractured) column. Regarding the characteristic of the heterogeneous column clean-up, initially the outlet solute concentration increased during the clean-up process and later it decreased until complete clean-up of the column was achieved. Further, the clean-up of the heterogeneous column was obtained for a larger number of pore volumes than that of the homogeneous column. In the next section we explain the reasons for the differences between the clean-up characteristics of columns made of homogeneous and heterogeneous porous media.

COMPARISON WITH OTHER PREVIOUS STUDIES

We found it appropriate to compare the experimental results of this study with those of some previous studies concerning entrapment of NAPL in heterogeneous porous matrices.

Geller and Hunt (1993) carried out experiments with a cylindrical horizontal column made of glass beads. In a central small cylindrical part of the column, NAPL was entrapped. Following the NAPL entrapment water was subject to flushing through the cylindrical domain. In the beginning a very small water discharge passed through the small central cylindrical part incorporating the entrapped NAPL due to its low water permeability. Under those conditions most of the water flow bypassed the domain part contaminated by entrapped NAPL. Therefore, the organic solute concentration at the outlet of the test section was initially very small. The clean-up of the central cylinder increased its water permeability and also the rate of entrapped NAPL solubilization and dissolution. Therefore, in initial stages, the solute concentration increased during the remediation process. Later, when the entrapped NAPL saturation became low it led to gradually decreasing solute concentration at the test section outlet. The experiments of Geller and Hunt (1993) were carried out with a homogeneous porous medium in which the contamination with entrapped NAPL was concentrated in a small cylindrical part of the test section. Such contamination led to heterogeneity of the contaminated domain. Therefore, as shown later, from the qualitative viewpoint the characteristics of remediation observed by Geller and Hunt (1993) were also typical of our study, but with results that were quantitatively different.

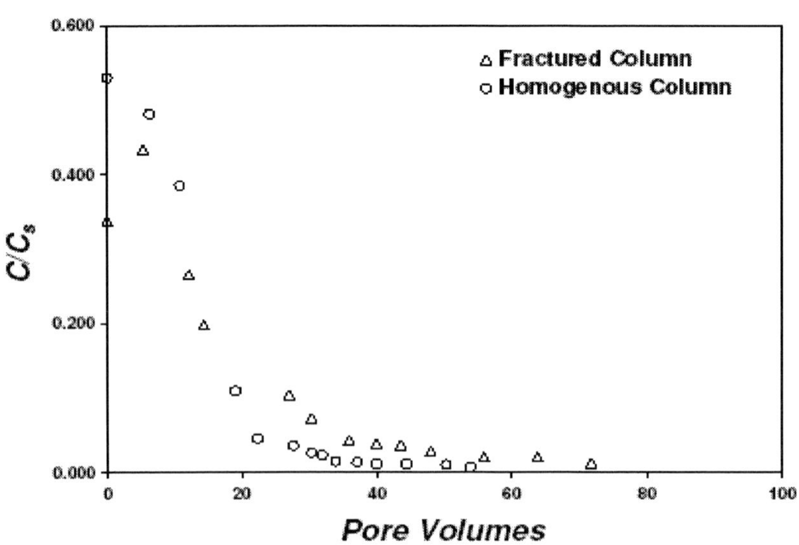

Figure 8. Changes of organic solute concentration at the top outlet of both columns made of Ashdod Sand.

The studies of Nambi and Powers (2000, 2003) were carried out with test sections consisting of columns made of fine sand free from NAPL into which lenses of coarse sand contaminated by entrapped NAPL have been inserted. During the experiments water was flushed through the test section. It seems that in initial stages most of the water discharge passed through the fine sand that was free from entrapped NAPL and bypassed the coarse sand whose water permeability was low due to the entrapped NAPL. Therefore, at that time the solute concentration at the test section outlet was low. Later, the increasing NAPL dissolution was accompanied with increasing water permeability of the coarse sand, leading to a gradual increase of the solute concentration at the outlet of the test section. And again in final stages the low entrapped NAPL saturation in the coarse sand lens led to a decreasing solute concentration at the test section outlet. The experiments of Nambi and Powers (2000, 2003) were carried out with a heterogeneous porous medium incorporating a major part of fine sand free from entrapped NAPL and a minor part (lens) of coarse sand contaminated by entrapped NAPL. The characteristics of clean-up of such a domain were qualitatively similar to those of cleaning-up the heterogeneous column used in this study. Therefore, in Figure 9 we compare our experimental results with those obtained by Geller and Hunt (1993) and those of Nambi and Powers (2003). According to Figure 9, we may conclude that qualitatively the experimental results of Geller and Hunt (1993) and those of Nambi and Powers (2003), as well as those of this study show similar phenomenon, namely, some initial water flow bypassing of regions contaminated by the entrapped NAPL, due to the low water permeability of those regions; and later the effect of bypassing diminishes because of the partial dissolution of the entrapped NAPL, which leads to increasing solute concentration at the outlet of the test section. In the final stages due to the low saturation of the entrapped NAPL the solute concentration gradually decreases at the test section outlet.

Unfortunately, from the two parameters governing the aquifer remediation process we could only measure with reasonable reliability the mobility number. However, the major parameter affecting the clean-up of the fractured permeable formation is the dimensionless interphase mass transfer coefficient (Rubin et al. 2008a). Therefore, we could only make some evaluations of our experiments and calculate how values of this parameter were subject to changes during the heterogeneous column remediation. However, such calculations might lead us to various types of speculations.

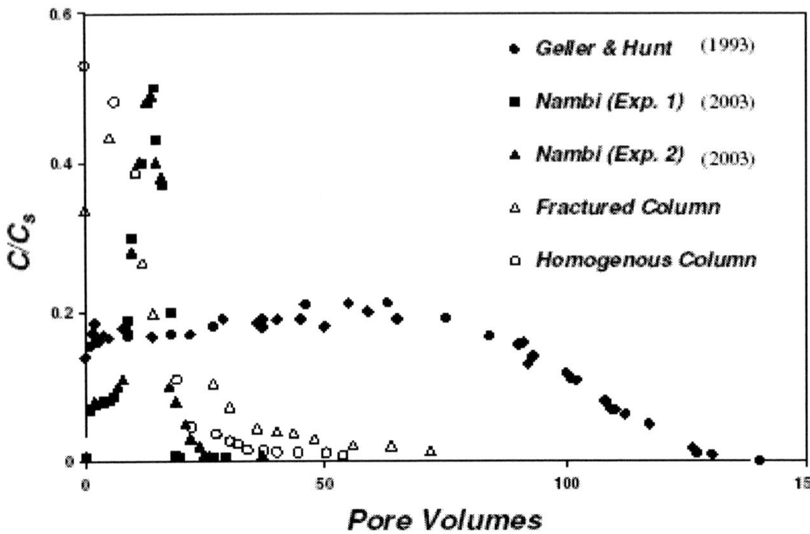

Figure 9. Comparison between the experimental results of this study with those of Geller and Hunt (1993) and Nambi and Powers (2003).

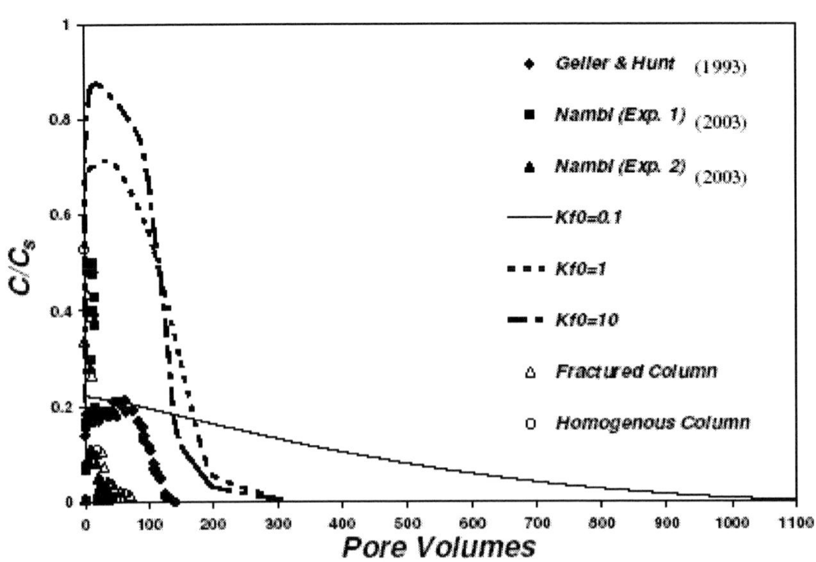

Figure 10. Comparison between the theoretical modeling and experimental results.

None of the experimental studies considered in the preceding paragraphs of Geller and Hunt (1993) and those of Nambi and Powers (2000, 2003), as well as this study may be represented by the fractured permeable formation modeling approach. However, qualitatively, and to some extent also quantitatively, the theoretical modeling approach predicts basic phenomena and characteristics typical of the effect of heterogeneity on the remediation of an aquifer composed of heterogeneous porous medium contaminated by entrapped NAPL. In Figure 10 we compare predictions provided by the theoretical model for three different values of the initial dimensionless interphase mass transfer coefficient with the experimental results obtained by the three experimental studies considered in preceding paragraphs. All results show identical qualitative phenomena, but there are significant quantitative differences between the various results. The presence of fractures causes bypassing of the groundwater flushing flow and reduces the effectiveness of the pump-and-treat remediation. Therefore, the presence of fractures leads to increasing times of remediation. A typical characteristic of pump-and-treat remediation of the heterogeneous aquifer is the increasing and later decreasing solute concentration at the outlet of the domain.

The phenomenon of by passing regions contaminated by entrapped NAPL should be taken into account whenever heterogeneity of the strata is concerned. As an example, Rubin and Narkis (2001) reported about the on site remediation of large quantities of loam soil contaminated by diesel fuel. The water permeability of loam soil is very low; therefore, this soil contaminated by diesel fuel was mixed with hay and arranged in reactors fed with air, water and nutrients. However, complete remediation of the loam could not be achieved, because the decayed hay created small channels through which the flashing fluid bypassed aggregates of loam contaminated by the entrapped NAPL. The conclusion of that study was that by artificially increasing the permeability of the NAPL contaminated porous medium while causing its heterogeneity could not enhance the remediation of the domain.

SUMMARY AND CONCLUSIONS

Decisions about contamination treatment are highly complex. This study examines parameters that may facilitate the decision process upon remediation of aquifers contaminated by LNAPL. Typical cases of aquifer contamination by LNAPL, namely fossil fuel, take place where the LNAPL is released close to the soil surface, infiltrates through the vadose zone, arrives at the water table and accumulates on top of the water table as LNAPL lenses. Later the LNAPL lenses spread horizontally; and due to seasonal and annual fluctuations of the water table the LNAPL penetrates into top layers of the free surface aquifer. In many cases the aquifer layers contaminated by the entrapped NAPL are composed of heterogeneous porous medium. In this study we have particularly referred to the example of contamination of a part of the CPA in Israel where the aquifer is comprised of a fractured permeable formation (fractured sand stone), namely permeable blocks in which fractures are embedded.

According to this study it is possible to classify fractured formations according to the value of the mobility number, which represents the ratio of the groundwater flow carried out through the permeable blocks to the groundwater flow carried out through the fracture network. If the mobility number is higher than 25 then the formation can be considered as homogeneous; if the mobility number is smaller than 0.01 then the aquifer is comprised of a fractured porous medium; and the aquifer comprises of fractured permeable formation provided the mobility numbers are in the range of 0.01 to 25.

Decisions concerning the remediation of the aquifer contaminated by entrapped NAPL are based on the proper balance among 4 domains: 1) Life and human health domain, 2) Environmental domain, 3) Water resources domain, and 4) Economic domain. The proper balance among these domains should take into account the option and partial involvement of natural attenuation, an appropriate time table of performing the remediation, applying hydraulic barriers, etc. The option of pump-and-treat procedure may be connected with full scale remediation, as well as with implementing hydraulic barriers, and allowing natural attenuation to be involved with the aquifer remediation. Therefore, in many cases, the pump-and-treat process seems to be at least part of the selected remediation alternative.

The experiments and theoretical analysis conducted in this study show that two major dimensionless parameters govern the remediation process, and therefore should be incorporated in systems associated decision support about remediation: 1) the mobility number that is connected with the preferential flow in the aquifer, and 2) the dimensionless interphase mass transfer coefficient that is connected with the interphase mass transfer from the entrapped NAPL into the groundwater flowing through the permeable blocks. The effect of the dimensionless interphase mass transfer coefficient is more significant than the effect of the mobility number on the remediation process.

By applying information about these parameters together with current data associated with the four domains, informed decisions about remediation can be made. For example, a decision support system can inform the decision maker about the cost of a pump-and-treat remediation, thereby considering the economic domain. The system can also assist the environmental domain by providing information about environmental standards and the tolerance to different levels of contamination. Similarly, the system can integrate information about the use of the contaminated water resource for drinking and other uses, as well as available alternative drinking water resources; thereby covering requirements about the life and human health domain, and those imposed by the water resource domain. By integrating such information with field information about the mobility number and the interphase mass transfer the decision support system can inform about the expected time and cost associated with pump-and-treat remediation, alternative actions that can be taken, and their impact on the four domains. The higher the interphase mass transfer and the higher the mobility number, the faster the contamination can be subject to clean-up, and therefore the lower the cost of complete remediation. It is up to the decision maker to decide whether pump-and-treat remediation should take place. Alternative actions, for example, may be to treat water to a low level of contamination for quality sufficient for irrigation (if needed) rather than complete treatment of the pumped contaminated groundwater. The decision support system can use the interphase mass coefficient and the mobility number to estimate the expected treatment time of the aquifer associated with arriving at different levels of contamination. Hence in using these parameters the system will be better equipped to provide information about the time, cost, health, and water resource outcomes associated with a partial treatment approach.

This study incorporates the development of a simplified theoretical 2D model describing the pump-and-treat remediation process of an aquifer comprising of a fractured permeable formation, which has been contaminated by entrapped NAPL. It is assumed that NAPL is washed out very quickly from the fracture network, and most efforts should be invested in cleaning-up the permeable blocks.

In order to evaluate the applicability of the theoretical model and the numerical results of the study, we have built an experimental set-up based on using two identical parallel columns with rectangular cross sections. One column was filled with homogeneous porous medium (glass beads, or pure fine sand) and the other was filled with the same homogeneous porous medium, in which a fracture model was constructed. The constructed fracture was made of mixtures of glass beads with pebbles.

Experiments with columns filled with glass beads were not conclusive, because the entrapped NAPL saturations were extremely low and NAPL saturation in the constructed fracture was of the same order of magnitude as that in the permeable blocks. Experiments in columns filled with fine sand were qualitatively in agreement with the theoretical results. However, we could not determine quantitatively all values of parameters involved in the remediation process. Further, our experiments only partly simulated the conditions represented by the theoretical model. We have also found some similarity between the results of this study and results of previous experimental studies. Typical phenomena of the pump-and-treat remediation of heterogeneous aquifers are:

1. In initial stages of the remediation the flashing water bypasses the NAPL contaminated regions (in which the water permeability is low) via regions free from entrapped NAPL. Under such conditions the organic solute concentration at the domain outlet is low and gradually increases with the gradual increase of the water permeability in regions contaminated by the entrapped NAPL, and
2. Later due to the partial remediation, the NAPL saturation becomes low leading to gradual decrease of the organic solute concentration at the outlet of the domain.
3. The fracture network causes bypassing of regions of low permeability (permeable blocks) of the domain and lead to increasing times of remediation by the pump-and-treat procedure.

In this study the fracture network represents the region free from entrapped NAPL and the permeable blocks have been contaminated by entrapped NAPL. Some previous studies reviewed in this manuscript incorporated other examples of heterogeneities, but the basic effects of the heterogeneity on the remediation of the contaminated porous medium were qualitatively similar.

REFERENCES

Birkhölzer, J., Rubin, H., Daniels, H., and Rouvè, G., 1993a. Contaminant transport in fractured permeable formation. 1. Parametric evaluation and analytical solution. *Journal of Hydrology* 144, 1-33.

Birkhölzer, J., Rubin, H., Daniels, H., and Rouvè, G., 1993b. Contaminant transport in fractured permeable formation. 1. Numerical solution. *Journal of Hydrology* 144, 35-58.

Geller, J.T., and Hunt, J.R., 1993. Mass transfer from nonaqueous phase organic liquids in water-saturated porous medium. *Water Resources Research* 29(4), 833-846

Kanfi, Y., 1986. Groundwater contamination by oil in the coastal plain aquifer of Israel. Ministry of Agriculture, *Water Commission,* Tel-Aviv.

Nambi, I.M., and Powers, S.E., 2000. NAPL dissolution in heterogeneous systems: an experimental study in simplified heterogeneous system, *Journal of Contaminant Hydrology* 44(2), 161-184.

Nambi, I.M., and Powers, S.E., 2003. Mass transfer correlations for nonaqueous phase liquid dissolution from regions with high initial saturations. *Water Resources Research* 39(2), 1030-1040.

Rubin, H., and Braester, C., 2000. *Field measurements, laboratory tests, and theoretical analysis of oil contamination in the Coastal Plain Aquifer of Israel (in Hebrew).* Department of Civil Engineering, Technion – Israel Institute of Technology, Haifa, Israel.

Rubin, H., and Narkis, N. 2001. Feasibility of site bioremediation of loam soil contaminated by diesel oil, *J. of Environmental Science and Health – Part A Toxic/Hazardous Substances & Environmental Engineering,,* A36(8), 1549 – 1558.

Rubin, H., Yaniv, S., Spiller, M., and Köngeter, J. 2008a. Parameters that control the cleanup of fractured permeable aquifers, *J. of Contaminant Hydrology,* 96(1-4), 128-149.

Rubin, H., Shoemaker, C.A., and Köngeter, J. 2008b. Screening of one-well hydraulic barrier design alternatives, *Ground Water*, 46(5), 743-754.

Rubin, H., Shoemaker, C.A., and Köngeter, J. 2009. Implementing a method of screening hydraulic barrier design alternatives, *Ground Water,* 47(2), 306-309.

Wiedemeier, T.H., Wilson, J.T., Kampbell, D.H., Miller, R.N., and Hansen, J.E. 1999. Technical protocol for implementing intrinsic remediation with long-term monitoring for natural attenuation of fuel contamination dissolved in groundwater, *Air Force Center for Environmental Excellence*

Technology Transfer Division Brooks Air Force Base, San Antonio, Texas.

In: Aquifers: Types, Impacts and Conservation ISBN: 978-1-61942-091-5
Editors: Z. Ouakili, et.al, pp. 33-58

Chapter 2

HYDROCHEMICAL FEATURES OF GROUNDWATER FROM AQUIFER SYSTEMS OCCURRING NEAR CENTRAL SÃO PAULO STATE, BRAZIL

Daniel Marcos Bonotto*[1], Luis Henrique Mancini**[2] and Érica Martini Tonetto[3]**

Departamento de Petrologia e Metalogenia, Universidade Estadual Paulista, Av. 24-A No.1515, C.P. 178, CEP 13506-900, Rio Claro, São Paulo, Brasil[1]

Instituto de Geociências, Universidade de Brasília, Campus Universitário Darcy Ribeiro, CEP 70910-900, Brasília, DF, Brasil[2]

Instituto de Geociências, Universidade Estadual de Campinas, Rua João Pandiá Calógeras No. 51, CEP 13083-870, Campinas, São Paulo, Brasil[3]

ABSTRACT

Brazil owns the highest availability of renewable hydrological resources in world. However, in the beginning of the 21st century, it was

* e-mail: danielbonotto@yahoo.com.br
** e-mail: luis.mancini@uol.com.br
*** e-mail: etonetto@rc.unesp.br

the 26th country in terms of social water availability as a consequence of the great unequal distribution of the hydrological resources in the country. This is because it is mainly concentrated in northern and portions of northeastern and central western regions, where the population density is low. This advanced degree of unequal water distribution has caused special attention to the groundwater resources due to problems related to the interaction between the society and the environment. São Paulo is the most populous Brazilian state, comprising ~ 40 million inhabitants distributed over 645 municipalities. It has the highest number of industries and economic production, reaching 31% of the Brazilian GDP-gross domestic product. Despite the vigorous industrial production that includes high technology goods, the state also is well developed in agriculture and cattle breeding. This advanced stage of agricultural and industrial growth causes to São Paulo State an accentuated use of groundwater resources in water-supply systems. Several lithologies of the Paraná sedimentary basin outcrop in São Paulo State. They consist on sandstones, conglomerates, diamictites, tillites, siltstones, mudstones, limestones, shales, silex, rythmites, basalts, diabases and different types of Cenozoic covers like the recent deposits, terrace sediments and the Rio Claro Formation (sandstones, conglomerate sandstones and muddy sandstones). Many of these rock types are important reservoirs for groundwater in São Paulo State. Rio Claro city is located at the northeastern edge of the Paraná basin, cropping-out several units of the basin in the region. Rio Claro city is the most important municipality in Corumbataí River basin that extends over an area of about 1581 km^2. The area considered in this chapter includes Rio Claro city and portions of Araras, Leme, Cordeirópolis, Santa Gertrudes, Itirapina, Corumbataí and Ipeúna municipalities. Different groundwater systems have been focused in this chapter, among them: a phreatic aquifer comprising a cover of unconsolidated materials from Rio Claro Formation; sediments from Tubarão Group, comprising a confined aquifer about 40 m thick; deep bodies of diabase that store water within their fractures. In general, sedimentary aquifers exhibit different flow and hydrochemical characteristics in relation to that of the fractured rock aquifers and these aspects will be considered in this chapter.

INTRODUCTION

The hydrologic cycle describes the continuous movement of water on, above, and below the surface of the Earth. The hydrologic cycle describes the processes that drive the movement of water throughout the hydrosphere, whilst a reservoir represents the water contained in different steps within the cycle.

Thus, the hydrosphere can be characterized as a system of different reservoirs from which water, solutes and energy are exchanged by the hydrological cycle (Vries, 2000).

The largest reservoir is the collection of oceans, accounting for 97% of the Earth's water (Vries, 2000). The next largest quantity (2%) is stored in solid form in the ice caps and glaciers (Vries, 2000). This small amount accounts for approximately 75% of all fresh water reserves on the planet (Vries, 2000). Most of the freshwater reserve actively involved in the hydrological cycle is found in the upper few kilometers of the subsurface. The subsurface consists of an upper zone with soil water and a lower saturated part with groundwater, where they are separated by the groundwater table (Vries, 2000). The depth at which groundwater is encountered depends on topography, subsurface structure and climate. Total storage of groundwater in a layer is determined by the porosity, but the ease at which groundwater can flow depends on a combination of interconnection and size of the pores (Vries, 2000).

The rock matrices exhibit variable properties to storage water, where sand and gravel or fractured bedrock often constitute good aquifer materials (Castany, 1967). In several areas, the main aquifers are typically unconsolidated alluvium, composed of horizontal layers of alternating coarse and fine materials deposited by water bodies (for instance, rivers and streams) (Castany, 1967). Fractures, faults, cracks and joints are also important rock properties that allow them to conduct and storage water.

The use of groundwater in São Paulo State, Brazil, has increased mainly in the last decade. It has been also expected a more accentuated increase in the next years, as a consequence of the technological advances of the sector coupled to the known advantageous of the groundwater in relation to superficial hydrological resources as, in principle, it is less subject do pollution. It has been calculated that approximately 60.5% of the urban nuclei of São Paulo State are totally or partially supplied by groundwater, as the exploitation of this resource is more intense at the western portion of the State (CETESB, 1998).

The exploitation of underground water resources in Rio Claro and adjacent municipalities near the center of São Paulo State, Brazil, has been increasing for industrial purposes, irrigation and human water supply. Rarely the water supply systems in the region are based exclusively in groundwater, occurring in some cities as a combination of surface and groundwater resources. Industrial and agricultural activities are very developed in the region and the important use of groundwater resources are for this objective, as well for drinking and recreation purposes.

The area focused in this chapter is an important center for the production of ceramics, and comprises portions of three different Management Units of Water Resources, classified according to heterogeneous physical characteristics, such as relief (Zaine and Penteado-Orellana, 1994), land use (Köffler, 1993), rock units (IPT, 1981), rain distribution (CEAPLA, 2001) and exploited aquifers (DAEE, 1981; Hirata et al., 1997).

Geological Aspects of the Parana Sedimentary Basin

The giant Paraná sedimentary basin constitutes a geotectonic unit established over the South American Platform since the Lower Devonian or Silurian (Almeida and Melo, 1981). It is located between parallels 10°-20° southern latitude and meridians 47°-64° western longitude. Its surface area corresponds to 1.6×10^6 km^2, comprising southern Brazil (1×10^6 km^2 in the states of Mato Grosso, Mato Grosso do Sul, Goiás, Minas Gerais, São Paulo, Paraná, Santa Catarina and Rio Grande do Sul), eastern Paraguay (0.1×10^6 km^2), NW Uruguay (0.1×10^6 km^2) and the northeastern extreme corner of Argentina (0.4×10^6 km^2) (Figure 1).

The accentuated basin subsidence, despite of oscillatory character, allowed an accumulation of thick sediments layers, basaltic lavas and diabase sills, where the total thickness of these deposits in the deepest portion of the basin reached up to 5 km (Gilboa et al., 1976).

The area where Paraná basin developed was much eroded in the Lower Devonian. Such surface had probably suffered marine transgression providing from the Andean domain. Despite the original extension occupied by this transgression is unknown, it has been assumed that the Paraná Group sediments constitute tracks of the erosion that affected the Devonian layers (Almeida and Melo, 1981). Paraná Group in São Paulo State is mainly characterized by whitish sandstones of mean to very coarse grain size with kaolinitic matrix. There are local intercalations of clayey to silty micaceous layers or fine sandstone. Conglomerates ~1-m thick also occur in the base, whose quartz and quartzite pebbles are disperse in a coarse sand matrix. There is also a generalized crossed stratification in benches relatively thick (1-3 m) which is originated by a water flow that also formed locally visible wavy marks.

Figure 1. Location of the Paraná sedimentary basin in South America.

Tubarão Group in Paraná basin contains records of the Permian-Carboniferous glaciation, marine ingressions, flora residues and coal layers. The paleoenvironment of these deposits exhibited great vertical and horizontal variation, causing difficulties on the Group sub-division, mainly when applied to large areas. Almeida and Melo (1981) considered the following sub-division in São Paulo State: Itararé, Aquidauana and Tatuí Formations. The Itararé Formation sediments predominantly constitute of fine to coarse sandstones, mudstones and diamictites, whose colors in the upper and lower parts are yellow, red and clear gray; the mean portion is represented by fine sandstones, siltstones and mudstones, whose colors are dark gray and yellow. The sandstones are frequently feldspatic or arkosian and form psamitic bodies, exhibit syngenetic sedimentary structures like wavy marks, crossed and gradational stratification, as well structures due to plastic deformation almost contemporary to deposition (DAEE, 1981). The Aquidauana Formation mainly constitutes of reddish sandstones and siltstones, also occurring red or greenish shales, conglomerates, diamictites and rythmites. The fine sandstones and siltstones constitute tabular or elongated bodies with intercalations of coarse feldspatic sandstone lenses or conglomerate lenses with sandy matrix and centimetric pebbles of rocks belonging to the crystalline basement. The non-massive sandstones exhibit plane-parallel or crossed stratification, layered or tangential. The Tatuí Formation is dominantly constituted by siltstones, also occurring sandstones, limestones, shales and silex layers. The array represents uniform sedimentation in contrast of the characteristic heterogeneity of the Itararé Formation (Almeida and Melo, 1981).

The Passa Dois Group sediments have been deposited in Upper Permian and are represented by (Almeida and Melo, 1981): siltstones, claystones and silty shales of clear to dark gray color, bituminous shales locally alternating with silicified fawnish limestones and restrict conglomeratic levels; essentially marine deposits including siltstones, shales and dark gray to black claystones with plane-parallel lamination; marine deposits comprising shales and dark gray to greenish/reddish finely laminated claystones alternating with siltstones and very fine sandstones in the presence of restrict lenses of oolitic limestones and silex; coastal plain deposits including very fine to mean greenish/reddish sandstones and subordinated claystones and reddish siltstones; possibly tidal plains marine deposits, including claystones, shales and gray, reddish or purplish siltstones with intercalations of carbonatic benches and layers of fine sandstones.

Pirambóia Formation is a sedimentary unit of expressive occurrence in central eastern São Paulo State, whose more characteristic morphological features constitute staggered scarps and extensive beaches existing over the tabular hills (Soares, 1975). The lower member presents clayey facies where dominates the plane-parallel and crossed slotted stratification of small size; the clay layers, sandy and silty shales are also frequent. The upper member is characterized by the presence of little to very clayey sandstone benches with plane-tangential crossed stratification of small to mean size, followed by very clayey sandstone benches with plane-parallel stratification, mudstones and sandy claystones in a clear cyclic repetition. The mud (silt + clay) content is much variable, with an average of about 20%; the grain size is homogeneous, varying between very fine to medium, with dominance of poorly selected fine sand. The Pirambóia Formation thickness is around 300 m in Tietê basin, reducing towards south and northeast of São Paulo State, where reaches 150 m in Analândia and 60 m in Franca (Soares, 1975). Soares (1975) situated the deposition period of the Pirambóia Formation between the Lower Triassic and Upper Jurassic, taking into account available fossils of chronological value.

The Botucatu Formation is almost entirely constituted by uniform sandstones exhibiting fine to medium grain size, well-selected rounded and frosted grains. They are reddish, possess crossed stratification of medium to large size and are friable or silicified. The Formation contains bodies of conglomeratic sandstones and conglomerate in the basal portion. The sedimentary package constitutes a genetic unit deposited in a desert environment mainly characterized by a monotonous succession of cuneiform bodies of regular to good selected sandstones, with mud (silt + clay) content generally lower than 10%. The Botucatu Formation thickness is much variable, but not exceeds 150 m in the outcrop area at São Paulo State. The average thickness is about 50-70 m and its fossil content in São Paulo State is very reduced and restricted to some crustacean, worm paths and vertebrate footprints (Almeida and Melo, 1981). Soares (1975) considered the Jurassic-Cretaceous as the upper age limit of the Botucatu Formation, suggesting the Jurassic as the lower limit.

The Serra Geral Formation constitutes a designation proposed by White (1908) to the group of basalts formed by flows through extensive fissure volcanism, estimated in 650,000 km^3, that are associated to intrusive bodies of the same composition, mainly constituting dykes and sills (Almeida and Melo, 1981). There is an area exhibiting several diabase dykes at northeast of São Paulo State, close to the border of the basin, where also occurs many sills. The flows are formed by dark gray to black colored afanitic rocks, with variable thickness (up to 50-100 meters) and horizontal extension exceeding 10 km. The central zone of the thicker flows is massif, microcrystalline, fractured by sub-vertical contraction joints that divide rock in columns. The upper portion of the flows contains vesicles and amigdales that are often horizontally elongated and contain the dominant percentage of the vitreous matter in the rock. The amigdales are partially or completely filled by chalcedony, quartz, calcite, zeolite and nontronite that is a mineral able to supply a greenish color. The basal zone of the flows exhibits similar aspects despite the abundance and thickness are slightly more reduced. Horizontal joints are frequent in the basal portions and top of the large flows that represent the laminar lava flow in the interior of the spills (Bagolini, 1971). In terms of petrographic features, the basalts of the Serra Geral Formation are primarily constituted by zoned labradorite associated to clinopyroxenes (augite and sometimes pigeonite), and secondarily by titanomagnetite, apatite, quartz and rare olivine and respective alteration products. The absence of Botucatu sandstone in the Serra Geral Formation has been suggested due to the difficulty of performing distinction between the flows and sills. However, Almeida and Melo (1981) pointed out several evidences of contemporary sedimentation and volcanism, indicating that the Botucatu Formation represents the various sub-environments of a large climatic desert of increasing dryness, whose existence prolonged up to the occurrence of the basaltic volcanism. The main phase of the magmatic manifestation occurred between 130 and 115 million years ago, as indicated the results obtained for samples dated by the potassium-argon method in whole rock (Amaral et al., 1966; Melfi, 1967). Additional estimate of the volcanism age by the ^{40}Ar-^{39}Ar method was performed by Renne et al. (1992) and Turner et al. (1994) that provided values between 137 and 127 Ma, which do not differ significantly of those reported by Amaral et al. (1966) and Melfi (1967). Turner et al. (1994) also suggested that the duration of the volcanism was 10 million years and estimated a low lava effusion rate corresponding to 0.1 km^3/year.

It has been verified an epirogenic uplift tendency in Brazilian territory inserted in the South American Platform when the Serra Geral Formation lava flows has finished. The northern portion of the Paraná basin has been characterized by located matting relatively to the remaining basin; Bauru Group accumulated in this depressed area in the Upper Cretaceous (Almeida and Melo, 1981). The Bauru Group sedimentation consisted on: fine to mean sandstones with well-rounded grains, purplish color, crossed stratification of medium to large size and local occurrence of cement and carbonatic nodules; very fine to medium badly-selected arkosian and massive sandstones locally presenting cement and carbonatic nodules; fine and very fine sandstones exhibiting cementation and carbonatic nodules with lenses of sandy siltstones and claystones occurring in massif benches, plane-parallel and crossed stratification of small to medium size; fine to gross sandstones comprising massive benches with tenuous crossed stratification of medium size and common presence of carbonatic nodules; sandstones with clayey cement and elongated lenses of shales; limonized conglomeratic sandstones, siltstones and oligomict conglomerate.

Table 1 summarizes the major stratigraphic units of the Paraná sedimentary basin, where widely occurs sandstones, conglomerates, diamictites, siltstones, shales, mudstones, limestones, basalts and diabases, among other rock types.

Some Physiographic and Climatic Aspects of the Parana Sedimentary Basin

The Paraná sedimentary basin encloses different types of climate due to its large size, i.e. there are significant latitude variations (about 20°), relief conditions (altitude ranging from 100 up to about 1800 m) and meteorological factors (Rebouças, 1980). The characteristic trace evidenced by the morphological scenario is its design in scaled highlands, as from east to west several scarps are crossed: Mar Mountain Range, Central Plateau and Serra Geral Western Plateau.

Rebouças (1980) pointed out four main hypsometric surfaces: 0-200 m, 200-500 m, 500-800 m and 800-1200 m. These surfaces were sculpted in sedimentary terrains of variable lithology and contain typical features of their substratum (structural platforms, scaled profiles, tabular cores, tables and correlated forms), remaining the characteristic of dissected plateaus and ending in smooth and outstanding *cuestas*. The hypsometric domain corresponding to 500-800 m, in its raised tabular form, conditions the formation of the expressive Serra Geral *cuesta*, related to the pronounced basaltic flows. The *Depressão Periférica* (Almeida, 1964) belongs to this surface, whose relief comprises several altitude variations and diversified lithology, corresponding to the outcrop zone of several sedimentary formations that constitute the confined aquifer systems in the basin. The sandstone plateau (Bauru Group) in the northern portion of the area is a highlight feature dissected by the river regressive erosion.

There is a large variation of vegetal cover in the ample domain of the Paraná sedimentary basin, including wide field areas intensively utilized for agriculture and livestock, areas covered by dense virgin forests or *cerrado* fields (Rebouças, 1980). The soils are generally fertile in the basin, well irrigated and drained by favorable hydrological system. The favorable climatic conditions allow the development of agricultural and pastoral activities in considerable extension and intensity. The uniformity of soil types is outstanding as the lateritic ones are dominant in the basin.

Rainfall is the major form of precipitation in Paraná sedimentary basin. The snow does not take an important role in the area. The hail is more frequently observed, but is locally important; it is a phenomenon verified in small shower portions and not frequently monitored by meteorological stations (Rebouças, 1980).

The isohyets map presented by Rebouças (1980) shows that the average annual rainfall (1931-1960) varies from 2400 to 1000 mm. The interval 1200-1400 mm virtually includes the whole inter-tropical domain of the sedimentary basin. The values higher than 1400 mm form concentric figures, reaching a maximum of 2400 mm in the mountainous domain of Paraná State and north of Santa Catarina State in Brazil.

Table 1. The stratigraphic column of the Paraná sedimentary basin since Upper Carboniferous up to Cretaceous. According to Almeida and Melo (1981).

ERA	PERIOD	GROUP	FORMATION	LITHOLOGY	FORMATION	LITHOLOGY
		Bauru	Marília	Fine to gross sandstones; lenses of siltstones, mudstones, and very fine sandstones; carbonatic nodules	Itaqueri	Sandstones with clayey cement; lenses of shales and conglomerates
			Adamantina	Fine/very fine sandstones; carbonatic nodules; lenses of sandy siltstones and mudstones		
	Cretaceous to Jurassic		Santo Anastácio	Fine/very fine sandstones; carbonatic nodules and cement		
Mesozoic			Caiuá	Fine fawnish sandstones; carbonatic nodules and cement		
		São Bento	Serra Geral	Gray to black basalts with interbedded fine sandstones; diabase sills		
			Botucatu	Reddish eolic fine sandstones; restrict layers of siltstones and mudstones		
	Triassic		Pirambóia	Reddish fine fluvial silt/clayey sandstones; layers of shales and variegated coloured clayey sandstones		

Table 1. (Continued)

Paleozoic	Upper Permian	Passa Dois	Rio do Rasto	Greenish to reddish very fine/fine sandstones; reddish mudstones and siltstones	Corumbataí	Gray/fawnish/reddish mudstones, shales, and siltstones with interbedded layers of fine sandstones
			Teresina	Dark gray/greenish/reddish shales and mudstones finely layered with alternating siltstones and very fine sandstones		
			Serra Alta	Dark gray/black siltstones, shales and mudstones		
			Irati	Clear to dark gray siltstones, mudstones, and shales, bituminous shales locally alternating with limestones and restrict conglomerate layers		
	Mean Permian to Upper Carboniferous	Tubarão	Palermo	Greenish gray siltstones; fine sandstones and conglomerates	Tatuí	Reddish/fawnish/greenish siltstones and fine sandstones; limestone concretions
			Rio Bonito	Siltstones and shales with carbonatic/clayey layers; very fine sandstones		
			Itararé	Arkosian sandstones, conglomerates, diamictites, tillites, siltstones, shales; restrict coal layers		
			Aquidauana	Mean to gross feldspatic reddish/fawnish sandstones; fine sandstones, conglomerates, siltstones, shales, diamictites		

Table 2. Results of chemical analysis of groundwater from different aquifer systems at Rio Claro city, São Paulo State, Brazil. According to Bonotto and Mancini (1992)

Sampling date	pH	Eh (mV)	O_2 (mg/L)	TOC[a] (mg/L)	Na (mg/L)	K (mg/L)	Ca (mg/L)	Mg (mg/L)	HCO_3^- (mg/L)	CO_3^{2-} (mg/L)	SO_4^{2-} (mg/L)	Cl^- (mg/L)	NO_3^- (mg/L)	F^- (mg/L)	PO_4^{3-} (mg/L)	TDS[b] (mg/L)
Rio Claro Formation (Well No. 1)																
10/19/1989	6.21	-	12.4	3.1	2.89	1.36	8.72	4.41	39.0	<0.1	<2.00	0.43	3.00	<0.010	<0.10	80
10/25/1989	6.00	-	12.4	3.2	2.75	1.36	8.77	4.43	38.0	<0.1	3.50	0.30	1.70	0.280	<0.10	120
11/01/1989	6.13	-	11.4	24.6	2.80	1.20	8.10	4.10	40.0	<0.1	<2.00	<0.10	1.40	0.085	<0.10	80
11/08/1989	6.46	-	12.0	27.8	2.40	1.20	8.30	4.10	37.0	<0.1	6.30	1.70	1.40	0.200	<0.10	120
11/16/1989	6.08	-	7.5	6.7	2.50	1.20	8.40	4.20	38.0	<0.1	<2.00	<0.10	1.40	<0.010	<0.10	120
11/23/1989	6.30	-	7.0	7.5	3.00	1.20	8.40	0.20	19.0	<0.1	<2.00	3.40	<1.60	0.085	0.10	160
11/29/1989	6.20	+149	6.5	3.4	2.50	1.10	8.40	0.20	39.0	<0.1	6.00	0.70	1.60	<0.010	0.10	80
12/06/1989	5.75	+173	11.0	2.8	2.50	1.20	8.50	0.50	39.0	<0.1	5.30	0.70	1.20	0.350	0.10	240
Fractured diabase (Well No. 2)																
10/19/1989	5.14	-	11.0	1.5	13.79	4.73	0.90	0.58	3.0	<0.1	<2.00	7.70	3.15	<0.010	<0.10	40
10/25/1989	5.20	-	11.0	2.1	14.31	4.97	0.93	0.64	2.5	<0.1	2.50	10.77	7.40	0.040	<0.10	80
11/01/1989	5.75	-	6.6	41.5	14.20	4.70	0.50	0.60	3.0	<0.1	7.10	8.40	7.90	<0.010	<0.10	80
11/08/1989	5.20	-	10.0	47.1	14.80	4.60	0.40	0.70	2.0	<0.1	<2.00	10.20	8.30	<0.010	<0.10	80
11/16/1989	5.25	-	9.5	5.5	12.50	4.60	0.40	0.70	3.0	<0.1	19.60	8.30	8.80	<0.010	<0.10	80
11/23/1989	-	-	4.5	5.2	15.10	5.00	0.50	0.70	<0.1	<0.1	2.00	14.40	8.80	<0.010	<0.10	120
11/29/1989	5.18	+212	5.8	2.1	15.50	5.00	0.50	0.20	4.0	<0.1	<2.00	8.20	8.30	<0.010	<0.10	-
12/06/1989	4.32	+248	8.5	2.1	15.00	5.40	0.40	0.20	3.0	<0.1	<2.00	8.60	8.70	0.085	<0.10	120
Tubarão Group (Well No. 3)																
10/19/1989	9.70	-	4.0	7.2	124.04	0.53	0.45	<0.10	191.0	34.0	39.75	3.07	<0.01	0.500	<0.10	320
10/25/1989	9.50	-	3.5	6.7	124.04	1.01	0.46	<0.10	188.0	33.0	39.00	3.28	<0.01	1.360	<0.10	320
11/01/1989	9.45	-	3.5	35.4	120.00	0.40	0.10	<0.10	189.0	35.0	54.50	5.30	<0.01	1.360	<0.10	320
11/08/1989	9.38	-	5.8	43.4	120.00	0.50	0.10	0.10	188.0	38.0	59.60	3.00	<0.01	1.670	<0.10	400
11/16/1989	9.02	-	3.7	9.5	90.10	0.50	0.10	<0.10	185.0	37.0	33.85	9.50	<0.02	1.570	<0.10	280
11/23/1989	9.00	-	2.5	9.2	111.0	0.50	0.20	0.20	129.0	29.0	41.00	11.60	<0.01	1.360	<0.10	360
11/29/1989	9.10	-30	2.2	5.8	125.00	0.50	0.20	0.20	186.0	37.0	42.10	3.00	<0.01	1.570	<0.10	360
12/06/1989	8.87	-20	3.4	6.4	127.00	0.50	0.20	0.20	188.0	37.0	40.00	2.60	<0.01	1.360	<0.10	400

[a]TOC = Total Organic Carbon; [b]TDS = Total Dissolved Solids.

HYDROGEOLOGICAL FEATURES OF THE PARANA SEDIMENTARY BASIN

The Botucatu aquifer is the most important aquifer system in the region, representing about 80% of the hydrogeological potential in Paraná basin. An aquifer system is defined as an array formed by one or more lithological and stratigraphic units possessing hydraulic continuity, hydrodynamic characteristics, recharge conditions, circulation, discharge, and water quality relatively similar. Under this view, Rebouças (1976) named Botucatu aquifer system to the lithological and stratigraphic units formed by Pirambóia and Botucatu Formations, however, Rebouças (1994) pointed out the need to revise the concept, considering them as two distinct aquifer systems due to the additional knowledge about their depositional models, facies, permeability and water quality, with outline to the fluoride content coupled to the lacustrine facies of Pirambóia Formation (Fraga, 1992). Such aquifer system deserved much attention since the late 1990s, receiving other names like Mercosul Giant Aquifer and Guarani Aquifer, after an informal talk among 3 geologists (Danilo Antón, Ernani Francisco da Rosa Filho, Jorge Montaño Xavier) as a reference to indians living in its occurrence area.

Guarani aquifer is confined in most of its extension and unconfined in the basin edges. The thickness increases towards the basin central portion, where the water potentiometric surface exhibits artesian conditions [the water table is 400 m northwards, reaching 1000 m southwards, as pointed out Rebouças (1988)]. The groundwater flow direction is from the phreatic exposed areas that surround the entire basin towards its central part, where there is a general southwestern resultant flow direction vector. Figure 2 illustrates the Guarani aquifer outcrop at the Paraná sedimentary basin, whereas Figure 3 shows the groundwater flow direction in Guarani aquifer at São Paulo State.

The relief influence is important on the characterization of the exploration conditions of the Paraná basin aquifers through tubular wells. The water table of phreatic aquifers becomes deeper in the highlands and mountains range domain, whereas may almost outcrop in the depressed relief zones. The confined aquifers, mainly Guarani aquifer system, exhibit artesian conditions in the hypsometric domains situated below about 500 m (Rebouças, 1976).

In terms of the exploration potential, other aquifer zones in the Paraná sedimentary basin include the Serra Geral Formation and Bauru Group (Rebouças, 1980). In general, aquifers inserted in Paraná, Tubarão and Passa-Dois Groups are mainly restricted to some zones in the outcrop domains, as they are difficult to exploit towards the basin center due to the great depths they are reached (Rebouças, 1980).

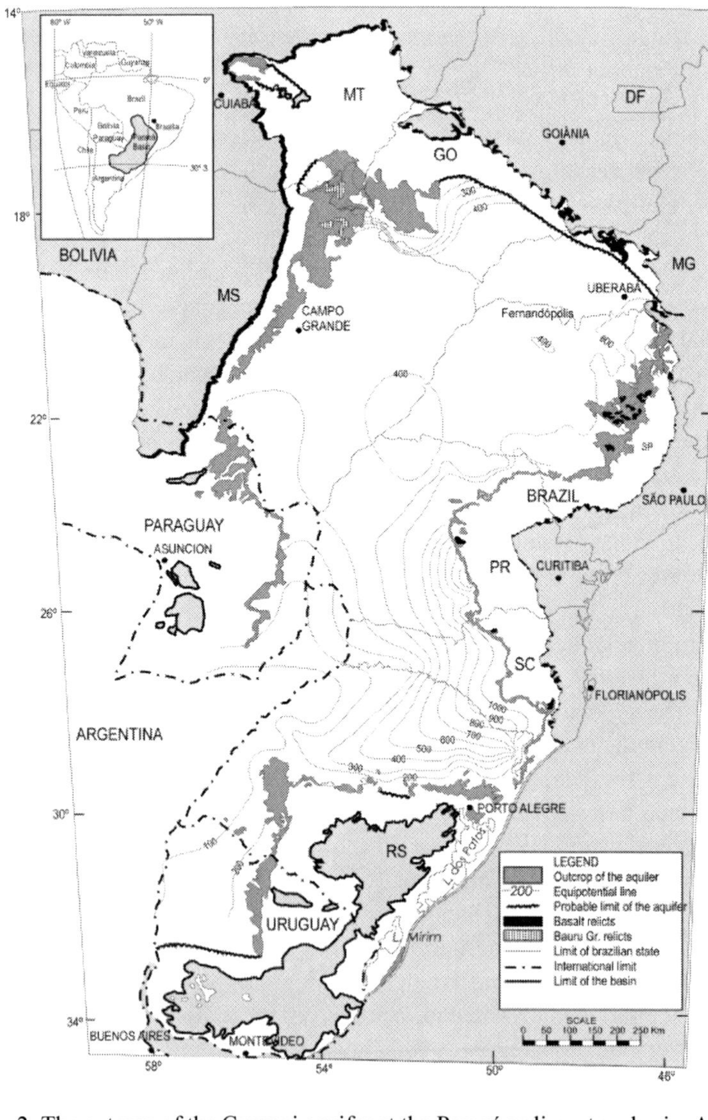

Figure 2. The outcrop of the Guarani aquifer at the Paraná sedimentary basin. Adapted from Araújo et al. (1995).

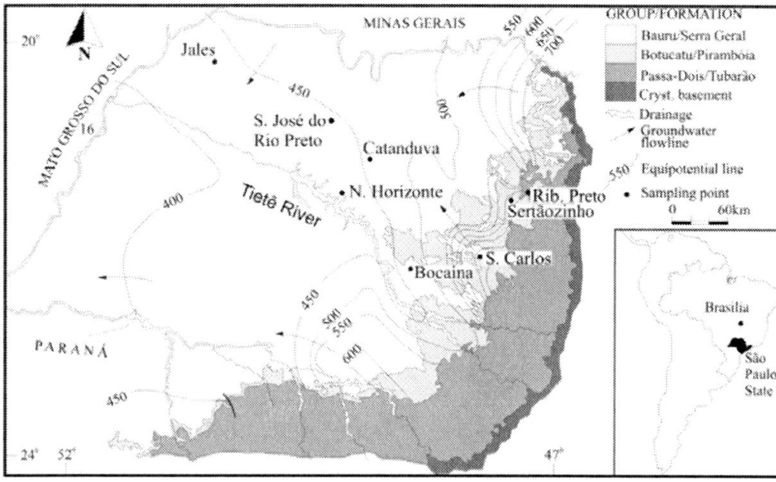

Figure 3. Simplified partial map of São Paulo State showing the Tietê River traversing the whole state, the main geological formations, and the groundwater flow direction in Guarani aquifer. Modified from Silva (1983).

The dominant granulometric characteristics of Botucatu Formation are of well selected and very fine sandstones accumulated by eolian processes under desertic conditions so that the aquifer system exhibit relatively homogeneous and isotropic hydrodynamic, porosity and permeability properties. On the other hand, the Pirambóia Formation sediments are composed of silty and shaly sandstones of fluvial-lacustrine origin that results in a large granulometric variation, i.e. there is dominance of mixing grain sizes (Rebouças, 1994).

Two outlets have been described in Guarani aquifer system (Gilboa et al., 1976): one allowing direct contact along the coast and the continental slope (a gap ~150 km wide in the crystalline framework in the states of Santa Catarina and Rio Grande do Sul in Brazil) and other masked by continental Quaternary deposits at SW end of the basin (NE Argentina). Some sporadic data indicate hydraulic conductivity values of 10^{-4}-10^{-5} m/s, effective porosity values of 10-20%, storage coefficient values of 10^{-3}-10^{-6} and average transmissivity of 10^{-3} m^2/s (Kimmelmann e Silva et al., 1989). The water is used extensively for drinking purposes, industrial activities and recreation (thermal swimming pools), with the yields of the wells varying from 10-150 m^3/h (phreatic exposed parts) up to 300-1000 m^3/h (confined artesian wells) (Gilboa et al., 1976).

A thick basaltic package of the Serra Geral formation (up to 1,500 m) overlies Guarani aquifer, reducing its exposed areas in non-continuous elongated strips, 10-100 km wide, along the edges of the proper Paraná basin, where the longest strip stretches between the states of São Paulo and Santa Catarina, in Brazil. Guarani aquifer overlies previous formations ranging from the igneous basement to the Paleozoic sediments of the Passa-Dois and Tubarão Groups, being covered by Cretaceous sediments of the Bauru-Caiuá formations.

Therefore, multiaquifer systems mainly comprising sandstones and basalts plus sediments from Passa Dois Group behaving as aquitards have been proposed to represent the hydrostratigraphy of the Paraná basin (Campos, 2000). The Passa Dois Group has often been considered an aquitard because of its major rock types, i.e. mudstones, siltstones and shales. However, structural lineaments, fractures in packages of fine grained sediments and the presence of carbonatic sandstone banks as well as extensive pumping allow water movement, and, consequently, this unit also behaves as an aquifer at many sites.

Groundwater in Paraná basin also occurs within the interflow zones and along cooling joints in basalts and diabases from Serra Geral Formation, where interbedded sediments greatly increase the average porosity of large volumes of rocks. The sandstones of Cretaceous age (Bauru Group) have an average thickness of 100 m, are moderately cemented, and exhibit adequate properties to storage water. The Paleozoic sediments (Devonian-Permian age) also provide water that is relatively mineralized in the central parts of the basin and contains H_2S in some wells.

Groundwater Resources at Rio Claro City and Neighborhood

The city of Rio Claro is situated at 22°23′S and 47°36′W, having an area of about 540 km^2 (Figure 4). It is located at the northeastern edge of the Paraná sedimentary basin, cropping-out several stratigraphical units of the basin in the region, for example: Itararé Sb-Group; Aquidauana and Tatuí from Tubarão Group; Irati and Corumbataí formations from Passa Dois Group; Pirambóia, Botucatu and Serra Geral formations from São Bento Group; intrusive rocks related to the lava flows from Serra Geral Formation and different types of

Cenozoic covers such as the Rio Claro Formation, where the urban area is situated.

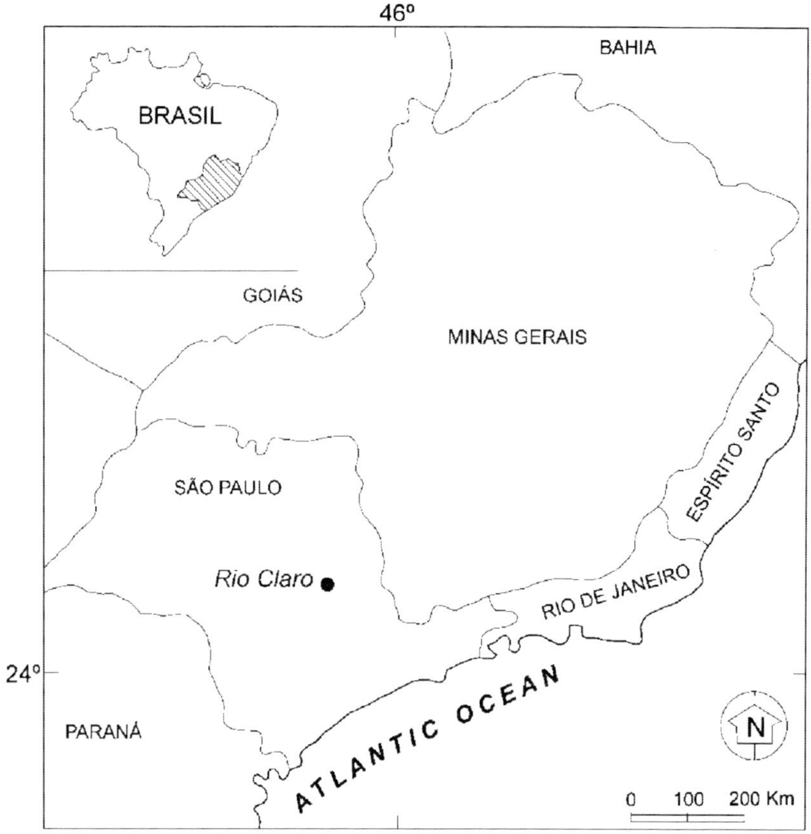

Figure 4. Location of Rio Claro city at southeast Brazil.

The climate of the region is Aw type (Köeppen classification), i.e. tropical rainy weather characterized by wet summer (October through March) and dry winter (April through September) (Inácio and Santos, 1988). The area often has 55-65 days of rain per year, with more than 80% of the precipitation falling between October through March (Bonotto and Mancini, 1992).

Rio Claro city is the most important municipality in Corumbataí River basin that extends over an area of about 1581 km^2 and includes Corumbataí River (extension of about 120 km) and tributaries (Claro stream, Passa Cinco River and Cabeças River). The Corumbataí river and affluents start flowing in the *cuesta* zone of the *Depressão Periférica* geomorphological province (Penteado, 1976), reaching Piracicaba river after crossing Rio Claro city. The area considered in this chapter includes Rio Claro city and portions of Araras, Leme, Cordeirópolis, Santa Gertrudes, Itirapina, Corumbataí and Ipeúna municipalities (Figure 5).

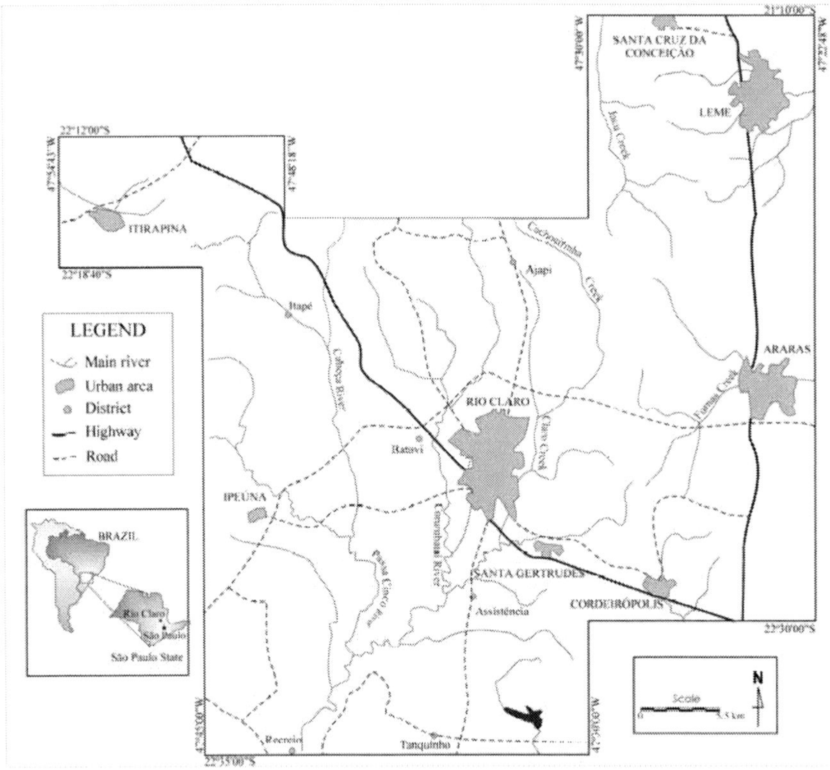

Figure 5. Location of Rio Claro city and other municipalities at São Paulo State, Brazil. According to Tonetto and Bonotto (2005).

A detailed hydrogeological characterization of Rio Claro and region is still a need, but the available information for other locations at the Paraná sedimentary basin provide useful insights on the investigated aquifers. According to the geological descriptions of the well profiles, there are a large number of units that can store and promote the circulation of water throughout the area, being verified as fissured (Serra Geral Formation) and porous (other formations) aquifers.

The Cenozoic covers like Rio Claro Formation constitute an important shallow unconfined local aquifer for the region, more subject to surface pollution (Oliva, 1999); impermeable sediments from Corumbataí Formation underlie this cover. A dug well for individual water supply intersected the water table of this aquifer at about 10-11 m in the beginning of the wet period; however, variations between 1 and 3 m have been identified due to seasonal variations (Cottas, 1983). The hydrochemical data for such well (No. 1) is given in Table 2.

Groundwater in Serra Geral Formation occurs within the interflow zones and along cooling joints in basalts and diabases, where interbedded sediments greatly increase the average porosity of large volumes of rocks. The water circulates predominantly through fractures and fissures in diabases at the area studied, what makes this aquifer very local, discontinuous, and heterogeneous. The diabases also occur as intercalated layers in some wells drilled at other formations (Botucatu, Pirambóia, Tatuí and Itararé). A 60-m deep well for supplying water to the rail company was drilled and intersected the water table at 30 m. The hydrochemical data for this well (No. 2) are shown in Table 2.

Water stored and circulating through porous sandstones from Pirambóia and Botucatu formations, representing Guarani aquifer at the area studied, has also been considered in this chapter. Major outcrops of these formations occur at northwest of the area studied (Figure 5). The sediments of Passa Dois Group at the area studied were cut by several drillings and some wells were sampled in this unit, despite their restriction for water storage and usage.

The Paleozoic sediments (Devonian-Permian age) generally provide water that is relatively mineralized in the more central parts of the Paraná sedimentary basin. Major outcrops of Tatuí and Itararé formations are verified at east and south of the area studied (Figure 5). The water storage and circulation occurs in sandy and conglomerate porous layers, but the presence of local impermeable fine-grained layers hinder the vertical movement of water. As observed elsewhere (DAEE, 1981; Hirata et al., 1997), the Tubarão aquifer is heterogeneous, discontinuous and anisothropic at the area under study. This aquifer occurs throughout the area, and the sampling was performed at several deep tubular wells, most of them drilled at Araras, Leme and near Rio Claro urban areas (Figure 5). Among them, it has been studied a 198.5 m deep well for supplying water to a club at Rio Claro city that was drilled in April 1970 and intersected the water table at 34 m. The hydrochemical data for this well (No. 3) are reported in Table 2. The sediments belonged to Tubarão Group (Tatuí Formation and upper layers from

Itararé Sub Group), and comprised a confined aquifer about 40 m thick, whose transmissivity was estimated to correspond to 12.3 m^2/h (Cottas, 1983).

HYDROCHEMISTRY IN THREE AQUIFER SYSTEMS AT RIO CLARO CITY

The hydrochemical data of the three aquifer systems systematically monitored at Rio Claro city are given in Table 2. It is clearly verified that the hydrochemistry is quite variable, as a consequence of the different aquifer rock matrices coupled to the systems investigated. The pH of Tubarão Group groundwater is much higher than that of the phreatic aquifer comprising the cover of unconsolidated materials from Rio Claro Formation and also of the groundwater occurring in diabase fractures. The waters providing from Rio Claro and Serra Geral formations exhibited Eh values from +149 to + 248 mV, thus, the Eh-pH diagram allows suggest they are circulating through a transitional environment, i.e. exhibits oxidizing to reducing conditions.

The groundwater exploiting Tubarão Group is very alkaline (pH between 8.87 and 9.70) due to the high content of dissolved sodium (between 90 and 127 mg/L) and the Eh potential redox ranged from -20 to -30 mV. Thus, the Eh-pH diagram indicates that this groundwater systemis circulating through a typical reducing environment that is diverse of that characterizing the waters providing from wells No. 1 and 2.

Several groundwater systems contain dissolved gases derived from natural sources. The gases involved in the normal groundwater geochemical cycles include: the atmospheric gases CO_2, O_2 and N_2; the gases derived from underground biochemical processes like the inflammable methane (CH_4), and hydrogen sulfide (H_2S). Although H_2S has not been quantified in Tubarão Group groundwater (well No. 3), it was readily identified in the samples collected due to its distinctive rotten-egg odor. Its presence was also confirmed by the lower dissolved O_2 content measured for these waters relatively to those providing from wells No. 1 and 2. It is compatible too with the reducing character attributed to this groundwater by the Eh-pH diagram (Brookins, 1988).

The hardness (HD) in water can be expressed as the equivalent of calcium carbonate by equation HD = 2.5 C + 4.1 M, where C and M are the contents of calcium and magnesium measured in milligrams per liter (Todd, 1959). All the values obtained for samples given in Table 2 are between 0 and 75 mg/L, so, the groundwaters are soft, according to the classification of Sawyer and McCarty (1967). These groundwaters are amenable to plotting on a standard Piper (1944) diagram because of the preponderance of typical anions and cations. So, the mapped facies are: calcium-bicarbonate for Rio Claro Formation, sodium-nitrate-chloride for diabase fractures and sodium-bicarbonate for Tubarão Group (Figure 6)

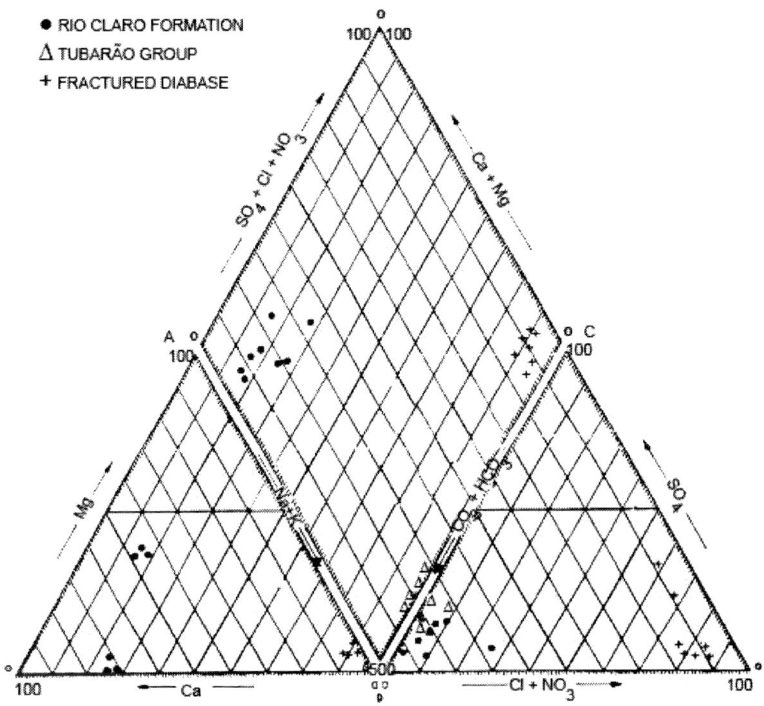

Figure 6. Chemical data reported in Table 2 plotted on a Piper (1944) diagram. According to Bonotto and Mancini (1992).

Regional Hydrochemical Relationships

Forty three groundwater samples were collected at the sites shown in Figure 5. The water temperature was always close to air temperature on the day of sampling, and all samples were clean, without color, smell or apparent turbidity. The suspended matter was less than 2 mg/L and the majority of the waters are distributed for human supply without previous filtration or treatment, sometimes chlorine and fluorine being added. The average values of chemical data obtained for waters of Serra Geral, Guarani and Tubarão aquifers indicate the sodium dominance in these three aquifer systems, and also the more accentuated presence of fluoride in waters of Tubarão aquifer.

The values obtained for alkalinity corresponded to the bicarbonate concentration since neither carbonate nor hydroxide were characterized possibly due to the low sensitivity of the analytical technique utilized, even for water samples of Tubarão aquifer whose pH values often exceeded 8. In a general way, water samples of Tubarão aquifer showed higher values for all parameters, following the same pattern observed for total dissolved solids, while groundwater of Guarani aquifer is more diluted.

Because several local and spatial differences were noticed, it was not possible to establish well-defined hydrochemical facies for each aquifer. Therefore, according to Piper (1944) diagrams, the majority of the waters were classified as mixed, consisting of a blend of waters with different chemical characteristics.

The water samples of Guarani aquifer exhibited significant statistical correlation for the following anions: bicarbonate and chloride (r=0.97); chloride and nitrate (r=0.97). They suggest local anthropogenic influences in the hydrochemistry of the aquifer as the wells are relatively shallow in the area studied (around 50 m depth) and they are also located in a recharge area/unconfined zone of the aquifer. According to the field observations, there are many agricultural activities taking place at this region, mainly sugar cane and orange plantations. It was also noticed a significant correlation between pH and Mg (r=0.88), indicating that the magnesium silicates dissolution contributes to the pH increase.

The Tubarão aquifer is widely exploited at the area shown in Figure 5 and, according to the Piper (1944) diagram, waters that leach the Paleozoic sediments are classified as mixed or are Na(K,Mg)-SO_4^{2-} dominated, probably due to the occurrence of mixing mainly with waters of Serra Geral and Passa Dois aquifers. Effective dissolution reactions are occurring at the strata representing the Tubarão aquifer as evidenced by the significant correlations found among several parameters, i.e. dry residue (~total dissolved solids) and sodium (r=0.71), sodium and sulfate (r=0.79), sodium and chloride (r=0.73), sodium and fluoride (r=0.65), sulfate and chloride (r=0.80), sulfate and fluoride (r=0.64), chloride and fluoride (r=0.78).

Considering the results for minor elements/compounds, except for aluminum and fluoride, the values did not exceed the limits established by

Brazilian legislation for potable waters (Agudo, 1987). Because clayey material Al-enriched is often utilized on the construction of the tubular wells, poor filtering systems in some cases may be responsible by the presence of aluminum concentration higher than the 0.2 mg/L limit for drinkable waters (Agudo, 1987). Fluoride concentration higher than the 1.5 mg/L limit for drinkable waters (Agudo, 1987) has been often reported in waters of Serra Geral and Tubarão aquifers at Paraná Sedimentary Basin (Fraga, 1992). Tannin and lignin were below the detection limit of the analytical technique (0.08 mg/L), suggesting low presence of organic matter in the groundwaters sampled.

Despite anthropogenic influences which may be affecting the presence of major anions in some of the groundwaters analyzed, it is important to refer that the developed industrial/agricultural activities and population concentration in urban areas apparently did not increase the concentration of minor elements and organic matter in deeper waters. This is very important, as among them there are heavy metals and dangerous contaminants responsible for serious damages to human health.

CONCLUSION

For many years, environmentalists, humanitarians, government agencies, politicians, educators, scientits and others have expressed the concern for protecting the underground resources against pollutants. The exploitation of underground water resources in Rio Claro and adjacent municipalities near the center of São Paulo State, Brazil, has been increasing as industrial and agricultural activities are very developed in the region and the important use of groundwater resources are for this objective, and also for drinking and recreation purposes. This chapter illustrated the use of some typical hydrochemical methods for characterizing dissolved constituents that are present in groundwater systems occurring in different geological contexts in the area studied. It has been reported the results of studies involving different aquifers (Cenozoic, Serra Geral, Guarani, Passa Dois and Tubarão) in the region, which consisted on the analysis of about 65 water samples from dug and deep tubular wells and of 2 spring water samples representative of infiltrating waters that predominantly leaches the Cenozoic (Rio Claro Formation) covers up to their contact with Paleozoic sediments of the Passa Dois Group. In general, the waters leaching Paleozoic sediments of Tubarão Group are more basic (pH often higher than 8) than those leaching Mesozoic sediments of Pirambóia and Botucatu formations ($pH_{average}$ = 5.9). Higher values for total dissolved solids (TDS) were generally obtained for groundwaters of Tubarão aquifer, while the waters of Guarani aquifer were more diluted and with less suspended matter, reflecting little minerals dissolution in sandstones during rock/water interaction. With exception of aluminum and fluoride, the values found for minor dissolved elements/compounds did not exceed the limits established by the Brazilian legislation for potable waters. The development of industrial/agricultural activities in the area studied and the population concentration in urban areas apparently did not increase the concentration of minor elements and organic matter in deeper waters.

ACKNOWLEDGEMENTS

FAPESP (Foundation Supporting Research in São Paulo State) and CNPq (National Council for Scientific and Technologic Development) in Brazil are greatly thanked for financial support of the investigations described in this chapter.

REFERENCES

Agudo, E. G. *Guia de coleta e preservação de amostras de água; CETESB* (Companhia de Tecnologia de Saneamento Ambiental): São Paulo, 1987.

Almeida, F. F. M. *Bol Inst Geogr Geol* 1964, 41, 169-263.

Almeida, F. F. M.; Melo, M. S. In *Mapa geológico do Estado de São Paulo; Pró-Minério; Ed.; PROMOCET:* São Paulo, 1981; v. 1, pp. 46-81.

Amaral, G.; Cordani, U. G.; Kawashita, K.; Reynolds, J. H. *Geochim Cosmochim Acta* 1966, 30, 159-189.

Araújo, L. M.; França, A. B.; Potter, P. E. *Giant Mercosul aquifer of Brazil, Argentina, Uruguay and Paraguay: hydrogeologic maps of Botucatu, Pirambóia, Rosário do Sul, Buena Vista, Misiones and Tacuarembó Formations;* Universidade Federal do Paraná: Curitiba, 1995.

Bagolini, B. *Mem Mus Tudentino di Sci Nat Ireuts* 1971, 19, 71-122.

Bonotto, D. M.; Mancini, L. H. *Geochim Brasil* 1992, 6 (2), 153-167.

Brookins, D. G. *Eh-pH diagrams for geochemistry*; Springer-Verlag: Berlin, 1988.

Campos, H. C. N. S. *Acta Geológica Leopoldensia* 2000, 23 (4), 1-50.

Castany, G. *Traité pratique des eaux souterraines*; Dunod: Paris, 1967.

CEAPLA (Centro de Análise e Planejamento Ambiental) 2001. *Atlas ambiental da Bacia do Rio Corumbataí* [http://www.rc.unesp.br/igce/ceapla/atlas/index.html].

CETESB (Companhia de Tecnologia de Saneamento Ambiental) *Relatório de qualidade das águas subterrâneas do estado de São Paulo 1997;* CETESB: São Paulo, 1998.

Cottas, L. R. *Estudos geológicos-geotécnicos aplicados ao planejamento urbano de Rio Claro (SP);* PhD Thesis; USP: São Paulo, 1983.

DAEE *(Departamento de Águas e Energia Elétrica) Estudo de águas subterrâneas - Região Administrativa 5 - Campinas; Secretaria de Obras e Meio Ambiente:* São Paulo, 1981; 639 pp.

Fraga, C. G. *Origem de fluoreto em águas subterrâneas dos sistemas aqüíferos Botucatu e Serra Geral da Bacia do Paraná;* PhD Thesis; USP: São Paulo, 1992.

Gilboa, Y.; Mero, F.; Mariano, I. B. *J Hydrol*, 1976, 29, 165-179.

Hirata, R. C. A.; Bastos, C. R. A.; Rocha, G. A. *Mapa de vulnerabilidade das águas subterrâneas no Estado de São Paulo*; CETESB (Companhia de Tecnologia de Saneamento Ambiental): São Paulo, 1997.

Inácio, A.; Santos, M. J. Z. *Bol Geografia Teorética* 1988, 18, 87-104.

IPT (Instituto de Pesquisas Tecnológicas) *Mapa geológico do Estado de São Paulo;* Monografias; IPT (Instituto de Pesquisas Tecnológicas): São Paulo, SP, 1981.

Kimmelmann e Silva, A. A.; Rebouças, A. C.; Santiago, M. M. F. *Radiocarbon* 1989, 31, 926-933.

Köffler, N. F. *Diagnóstico do uso de solos agrícolas da bacia do Rio Corumbataí, São Paulo;* IGCE-UNESP: Rio Claro, 1993.

Melfi, A. J. *Geochim Cosmochim Acta* 1967, 31, 1079-1089.

Oliva, A. *Caracterização hidroquímica de águas subterrâneas no município de Rio Claro (SP);* Graduation monography; IGCE-UNESP: Rio Claro, 1999.

Penteado, M. M. *Geomorfologia do setor central-ocidental da Depressão Periférica no Estado de São Paulo;* Série Teses e Monografias, 22; Universidade de São Paulo: São Paulo, SP, 1976.

Piper, A. M. *Trans Amer Geophysical Union* 1944, 25, 914-928.

Rebouças, A. C. *Recursos hídricos subterrâneos da Bacia do Paraná – Análise de pré-viabilidade;* Post PhD Thesis; USP: São Paulo, 1976.

Rebouças, A. C. In *Anais I Congresso Brasileiro de Águas Subterrâneas;* ABAS (Associação Brasileira de Águas Subterrâneas); Ed.; ABAS: Recife, 1980; pp. 35-47.

Rebouças, A. C. *Episodes* 1988, 11, 209-214.

Rebouças, A. C. In *Anais VIII Congresso Brasileiro de Águas Subterrâneas; ABAS (Associação Brasileira de Águas Subterrâneas); Ed.;* ABAS: Recife, 1994; pp 500-509.

Renne, P. R.; Ernesto, M.; Pacca, I. G.; Coe, R. S.; Glen, J. M.; Prévot, M.; Perrin, M. *Science* 1992, 258, 975-979.

Sawyer, C. N.; McCarty, P. L. *Chemistry for sanitary engineers*; McGraw-Hill: New York, 1967.

Silva, R. B. G. *Estudo hidroquímico e isotópico das águas subterrâneas do aqüífero Botucatu no Estado de São Paulo*; PhD Thesis; USP: São Paulo, 1983.

Soares, P. C. *Rev Bras Geoc* 1975, 5, 229-251.

Todd, D. K. *Ground Water Hydrology*; John Wiley & Sons, Inc: New York, 1959; 320 pp.

Tonetto, É. M.; Bonotto, D. M. *Environ Geol* 2005, 47 (7), 942-955.

Turner, S.; Regelous, M.; Kelley, S.; Hawkesworth, C. J.; Mantovani, M. S. M. *Earth Planet Sci Lett* 1994, 121, 333-348.

Vries, J.J. de In *Environmental isotopes in the hydrological cycle;* Mook, W. G.; Ed.; *UNESCO-IAEA*: Paris-Vienna, 2000; International Hydrological Programme, IHP-V, Technical Documents in Hydrology, 39 (1).

White, I. D. *Relatório final da Comissão de Estudos das Minas de Carvão de Pedra do Brasil;* Imprensa Nacional: Rio de Janeiro, 1908.

Zaine, J. E.; Penteado-Orellana, M. M. In *Anais XXXVIII Congresso Brasileiro de Geologia;* SBG (Sociedade Brasileira de Geologia); Ed.; SBG: Camboriú, 1994; Vol. 1, pp. 441-447.

In: Aquifers: Types, Impacts and Conservation ISBN: 978-1-61942-091-5
Editors: Z. Ouakili, et.al, pp. 59-84

Chapter 3

AQUIFER SYSTEM CHARACTERIZATION USING INTEGRATED GEOPHYSICAL METHODS

***Boutheina Farhat*[1], *Ismail Chenini*[1,*], *Abdallah Ben Mammou*[1] *and Sfar. Felfoul*[2]**
[1]Department of Geology, Faculty of Sciences of Tunis, 2092 Tunis El Manar University, Tunisia
[2]Agronomic National Institute of Tunis, 43, Charles-Nicolle Avenue, 1082 Tunis, Tunisia

ABSTRACT

The assessment of the hydrogeological conditions of a multilayered aquifer system requires a good knowledge of the aquifer geometry. The aquifer characterization is essential for the management of groundwater resources. This study has been carried out in the Mornag plain in North East of Tunisia and focuses on the geophysical methods application. It is a favourable site for assessing the performance and suitability of the geophysical methods for aquifer system identification and groundwater exploration. Thus, two complementary geophysical investigations comprising electrical resistivity and geophysical log analysis were carried out. This study provides a complete electrical image of the underground.

[*] Author to whom all correspondence should be addressed: E-mail: chenini_ismail@yahoo.fr; Phone: +216 97282335; Fax: +216 70860325

Results show that the prospected area is characterized by the succession of several levels with contrasted resistivities, which are often affected by faults. This study should be useful for choosing the best sites for reconnaissance borings or test wells that will precede the exploitation of the aquifer system, in the future.

Keywords: aquifer geometry; VES; Geophysical logging; groundwater exploration

1. Introduction and Objectives

Groundwater is not only an important component of water resources but also a reliable source of fresh water for a variety of purposes including domestic, industrial and irrigational uses. Nowadays, with increasing of population and life standards, there is a growing need for the utilization of groundwater resources. Indeed, due to some natural and anthropogenic causes such as demographic growth, unplanned urbanization and economic development (Jones 1999; Delmine et al. 2007; Perry et al. 2009) the quantity and the quality of groundwater resources continues to decrease (Custodio et al. 1987; Vengosh and Rosenthal 1994; Grassi et al. 2005; Ayvaz 2009; Raidla et al. 2009). Therefore, sustainable management strategies should be developed by decision makers to optimally utilize the groundwater resources. The characterization of the aquifer system is essential for effective water resource management. This can be improved by hydrogeological and complementary geophysical surveys. Such an integrated approach is developed in the important aquifer system of Mornag.

Mornag area is an important agricultural district of Tunisia, and groundwater has been overexploited since several tens of years (Horriche 2004). Therefore, a strong piezometric depression and salinization of shallow aquifer were distinguished (Lassoued et al. 1995). As a consequence of this degradation, water is now mainly pumped up from a deeper aquifer, and artificially injected in the aquifer from the Mejerda-Cap Bon canal is. Despite the importance of the Mornag aquifer, this is the first study devoted to the understanding of its geometry.

Numerous papers illustrate the utility of “indirect” approaches (geological or geophysical) to complement direct hydrogeological methods in defining subsurface heterogeneity in aquifer properties (Kelly et al. 1993; Crumbling et al. 2003; Rubin and Hubbard 2005). However, geophysical methods may provide a relatively low-cost approach to define aquifer characteristics (Kelly et al. 1993). Several geophysical methods have been employed with varying degrees of success to characterize a wide variety of aquifer types (Bowling et al. 2007).

Classical geophysical methods are widely used including vertical Electrical Soundings (VES) (Ritzi and Andolsek 1992; Ebraheem et al. 1997; Essahlaoui 2001; Shaaban 2001; Pham et al. 2002; Asfahani 2007), electromagnetic (Slater et al. 1998), Magnetic Resonance Sounding (MRS) (Schirov et al., 1991; Legchenko et al., 2002; Legchenko et al., 2004; Lubczynski, 2004), seismic reflection and/or refraction (Steeples and Miller 1990; Bradford et al. 1998; Cardimona et al. 1998; Kim et al. 2004; Harry et al. 2005), 3D seismic (Giustiniani et al. 2009), gravity, Radar data GPR and radioactivity measurements (Vogelsang 1995). Some geophysical data interpretations have an ambiguity due to the limitations of each method and/or site dependence and are unable to fully oruniquely describe subsurface hydrogeologic properties.

Thus, the combination of several geophysical methods allows a much more detailed and reliable assessment of aquifers than what was possible with only one geophysical method. Moreover, the uncertainty of geophysical interpretation can be notably reduced when several methods are jointly used (Goldman et al. 1994; Albouy et al. 2001).

There are numerous studies on the hydrogeological characterization using the combination of several geophysical methods (Savadogo 1997; Shtivelman et al. 2000; Vouillamoz et al. 2002; Yaramanci et al. 2002; Krivochieva 2003; Gouasmia et al. 2006; Sumanovac 2006; Tabbagh 2006; Bowling et al. 2007; Vouillamoz et al. 2007; Kim et al. 2008; Wattanasen et al. 2008).

In this paper, two complementary geophysical investigations have been carried out VES and geophysical logs. These data are also used in conjunction with boreholes data for an integrated interpretation. The main target of this study is primarily to improve the knowledge of the geometry and characteristics of the aquifer system in depth and also to obtain a new structural and hydrogeological model for the Mornag plain.

2. Geological and Hydrogeological Settings

The Mornag plain is located 20 km South-East of Tunis extending over a surface of 250 Km^2 (Figure 1).

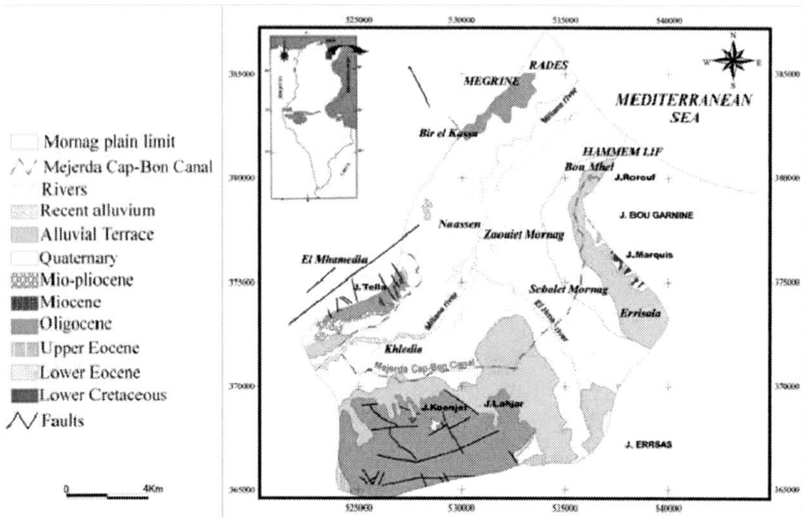

Figure 1. Location and geological map of the survey area.

The Miliane river has a permanent flow and is the main collector of surface waters in the basin. Its main affluent is the El Hma river. The geology of the studied area is characterized by a thick succession of mainly Cenozoic and Quaternary deposits (Figure 1). Triassic and Jurassic sediments occur at the bottom and are mainly composed of clays, gypsums, sandstones, dolomitic limestones and breaches. The Eocene deposits are essentially composed of limestone, attributed to the Lower Ypresian-Lutetian, and two levels of clays with small marl-limestone intercalations, separated by a coquina sandstone bed, corresponding to the Lutetian. The Oligocene sediments essentially occurring in the western part of the studied area (Khledia) comprises sandstones and limestones. The Miocene series correspond to marls and, to a lesser extent, sandstones. In the central part of the plain, a thick fluvial continental Plio-Quaternary series outcrop which are generally composed of sand, silt, clay, sandstone and conglomerate, resulting from the dismantling of the surrounding relief (Ladjmi 1960). The Quaternary series outcrop over a wide area of the plain as a thick sandy clay facies, sands with few gravels, pebbles and silts.

According to Ennabli (1977) the Mornag plain contains a multi-layered aquifer. The deep reservoir is found in the detrital Oligo-Miocene formations that crop out in the South-Western part of the plain (Ennabli 1980). Alluvial deposits (predominantly interbedded sandstones, sands, conglomerates and siltstones) are developed at the top of the upper aquifer and generally extend over all the plain. This phreatic aquifer is separated from the deep aquifer by a thin clayey formation, which may contain lenses of a conglomeratic sand from the surrounding landforms. Locally, the aquifers can be in hydraulical connection, due to the local morphology and tectonics, and can be considered as a unique aquifer system within a common hydrogeological basin. The degree of the hydraulic discontinuity between the phreatic aquifer and the lower confined one depends of the thickness and lithology of the confining units between them.

3. Geophysical Surveys

Two geophysical methods have been applied, in order to characterize the spatial distribution of the hydrogeologic facies thickness and the sediments nature to achieve a better knowledge of the aquifer geometry. The first method deals with the analysis borehole data (lithological logs, natural gamma activity and electrical resistivity logs). The second suggested approach is based, essentially, on the electric data interpretation (VES). A total of 134 geophysical logs from available boreholes were used. The borehole measurements consisted of natural gamma activity and shallow and deep (long and short) normal electrical resistivity. Analysis of gamma and electrical resistivity logs allowed the detection of the aquifer and the determination of the aquifer geometry. However, intervals of very low gamma activity and relatively high resistivities clearly distinguish the permeable unities composed of high resistive coarse-grained soils, from the impermeable relatively low resistive lithologies composed mainly of clays. The gamma and electrical resistivity logs in combination proved to be particularly effective in correcting description of lithological logs made by examining cuttings. The VES's are applied, to understand more precisely the electrical characteristics of the studied geological formations both in lateral and vertical directions. During the interpretation of the geoelectrical results, particular attention is given to the subsurface tectonics. A total of 230 VES's using the Wenner array were carried out in the study area (Astier 1971; Dobrin 1976; Koefoed 1979; Parasnis 1997), with a maximum current electrode half-spacing (AB /2) of 400 m, from which an investigation depth of about 100 m could be expected and would be less where conductive clay at relatively shallow depths is presented; so that only Mio-Plio-Quaternary shallow aquifers could be detected. Data was interpreted using the conventional curve matching and computer iteration method. Most of the VES's are conducted close to available water boreholes to correlate the data of the boreholes with the resultant resistivity data. Unfortunately, the investigation depth of this method is relatively low, about 1/4-1/10 of the distance between the current electrodes.

For deeper exploration, the geophysical and lithologic logs analysis are more suitable. In fact, these two methods are quite complementary, and the use of the combined Geophysical logs/VES methods can provide a more complete electrical image of the Mio-Plio-Quaternary sedimentary series.

4. Results and Discussion

4.1. Processing of Geophysical Logs Data

Detailed characterization of aquifer systems can be estimated by the interpretation of geophysical and lithologic logs from boreholes. However, results allow a much detailed and reliable assessment of aquifers and also the estimation of the water quality, transimissivity, clay content, grain size distribution and ground water potential of the subsurface formations. The results of such interpretation help in the calibration of the electrical sounding.

The results reveal that the prospected zone is characterized by the succession of several levels with heterogeneous lithology. The Mio-Plio-Quaternary is known for its high aquifer potential which consists of sand, gravely sand, sandy clay and clay intercalations. Therefore, the heterogeneities of these lithological units could be attributed to two main factors: areal variations in the grain size distribution and the gravel-sand-silt-clay ratios. On the whole, the Mio-Plio-Quaternary sediments are composed of a multilayered aquifer system with an alternation of highly conductive layers, attributable to a permeable sequence, and low resistivity layers corresponding to an impermeable sequence.

The interpretation data of the majority of boreholes allows to define three main layers A, B and C (Figure 2). The first layer corresponds to the top soil whose lithological nature correspond to a clayey-sandy sequence. The second layer consists of medium to coarse grained sand with a little clay content. This layer could present a hydrogeological interest. The third layer is composed of fine grain sediments with dominant clay. To understand more precisely the spatial distribution of the aquifer thickness and the sediment nature, fourteen correlations across the study site constructed from natural gamma logs and electrical resistivity logs were established (Figure 3).

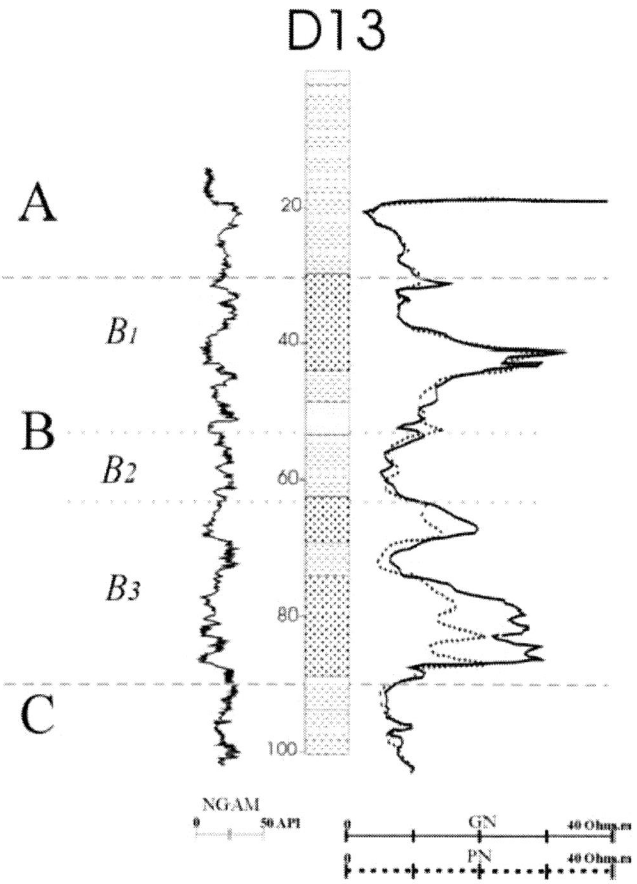

Figure 2. Geophysical log composite of natural gamma, deep and shallow resistivity (long and short normal) obtained in borehole D13.

4.1.1. Geophysical Cross Sections along the Profile C5

The examination of this profile shows strong lateral and vertical variations in the lithologic characteristics of the lithostratigraphic units (Figure 4). Three main layers dominated by coarse materials were defined (R1, R2 and R3) which could present a hydrogeological interest and it indicates a heterogeneous lithology composed mainly of sandy sequence. These layers were separated by two clay intercalations (M1 and M2).

The surface layer consists of medium to coarse grained sand and gravel with clayey intercalations.

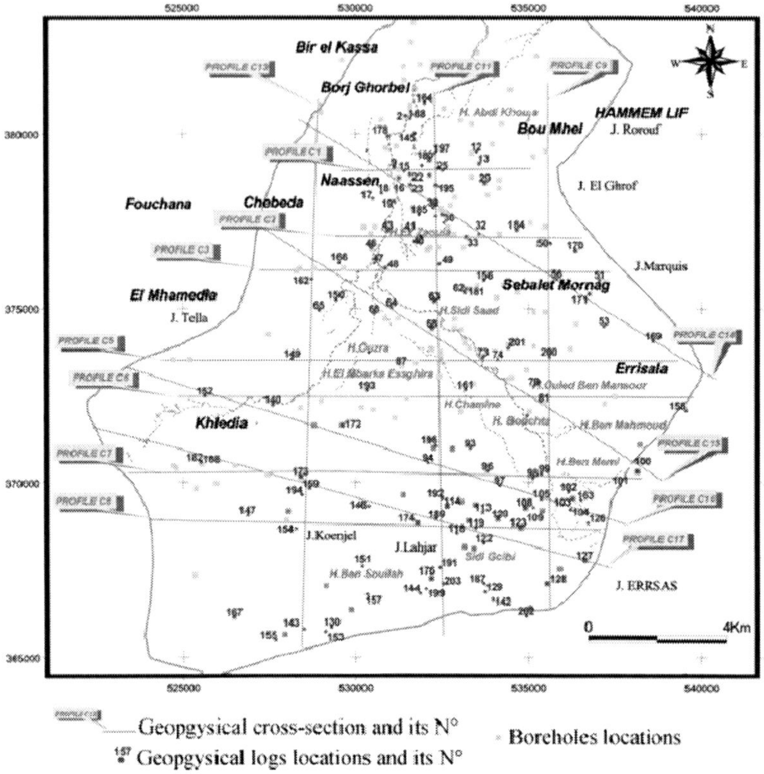

Figure 3. Location of the boreholes and the cross-section.

The average depth of this layer is between 10-40m which constitutes a phreatic aquifer (R1). The second layer consists of a permeable sequence, ranging from 20-30m in thickness. It is composed of coarse-grained material, and corresponds to the main aquifer (R2). The third layer corresponds to a semi deep aquifer (R3) with thickness varying from 30 to 35m and extends to a depth of about 130m. It is essentially composed of coarse grained sand and gravels. This layer thins and shows lateral facies variations westward of the study area. However, it changes from permeable, highly resistive sediments, to relatively low resistive lithologies. The two impermeable layers (M1 and M2) could decrease the mutual vertical relations by drainance between the three aquifer levels.

These layers thins (e.g. at D74, Figure 4a) or are occasionally absent (e.g. at D169, Figure 4a), occurring as a result a continuity and a direct communication between the three aquifers.

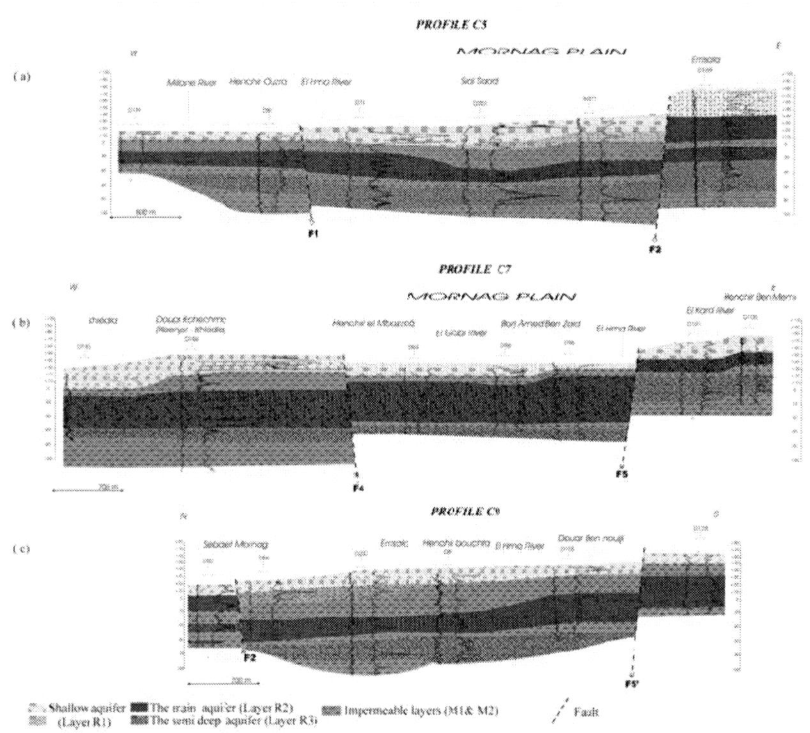

Figure 4. Geophysical cross sections (a) along the profile C5, (b) along the profile C7 and (c) along the profile C9.

From a structural point of view, the Mio-Plio-Quaternary replenishment is affected by two major faults (F1 and F2) promoting a manifest subsidence of the sedimentary section in the central part of the area (Figure 4a). This structural pattern explains the thickening of the deposits in this trough. The vertical displacement of the fault F2 is important, which is justified by the well developed communication between the aquifers.

This arrangement can play an important role in the exchange and the recharge of the aquifers that may act as a hydrogeological relay. In spite of the importance of this fault, the three aquifer systems are hydraulically connected to each other making the major aquifer system of the study area.

4.1.2. Geophysical Cross Sections along the Profile C7

This cross section presents the same lithological units as described in the previous section. It reveals the thickening of the aquifer layers to the profit of the impermeable levels thinning (Figure 4b). These clayey levels present some specificities: (1) They can thins (e.g. at D 182), particularly in the border area near Oligo-Miocene outcrops, or is occasionally absent (e.g. at D100 and D101) giving rise to the first (superficial] aquifer resting directly on the main aquifer, especially in the East of the study area. (2) Locally, they can disappear (e.g. at D182 and D159) giving rise to the main aquifer resting directly on the semi-deep aquifer, in the western corner of the area.

Examination of this section shows two major faults (F4 and F5), that led to the strong lateral and vertical variations in the thickness and depths of the aquifer levels. The hydrogeological characteristics and the hydrodynamic behaviour of the aquifer appear to be strongly controlled by these faults. This is justified by the well developed communication between these aquifers in fact; F4 and F5 have not played the role of hydraulic thresholds. Generally speaking, we deduce that the underground water resources are contained in three adjacent or superimposed aquifers which have unequal extent and importance.

4.1.3. Geophysical Cross Sections along the Profile C9

The faults F2 and F5', identified in this profile, show the existence of a graben structure, which would explain the thickening of the Mio-Plio-Quaternary series (Figure 4 c). In the remaining area, the aquifer system becomes generally shallow and thin. The structural pattern influences not only the variations of the aquifer's depth and thickness, but it has also a direct effect on the hydrological characteristics, especially in the northern and southern parts of the area.

4.2. Geoelectrical Resistivity Survey

The electrical resistivity of rocks depends on many factors such as groundwater salinity, saturation, aquifer lithology and porosity (Kelly 1977; Heigold et al. 1979; McNeill 1980; Kelly and Frohlich 1985; Mazac et al. 1988; Ahmed et al. 1988). There are no general guidelines for interpreting lithology of the contrasting resistivity layers, as each area has its own conditions. Accordingly, interpretation must be based on local databases that include information on the lithological and hydrogeological nature of the investigated area; these are acquired from boreholes.

The already lithological logs of drilled boreholes in the study area, made only from cuttings examination, could not be a source of information for calibration purposes, since their lithological description is not precise. For this reason, several geophysical logs (gamma and electrical resistivity), located near VES, are selected to correct the description of these lithological logs. The results of such calibration provide a complete electrical model of the underground and helped in the construction of the true resistivity spectrum of the sedimentary series.

The interpretation of the resistivity measurements, calibrated from surface geology and borehole data (geophysical and lithological logs), shows that the prospected zone is characterized by the succession of, at least, five levels with contrasted true resistivities, as will be shown later on the transversal and longitudinal profiles, from top to bottom as follows: A superficial layer, generally thin and highly heterogeneous.

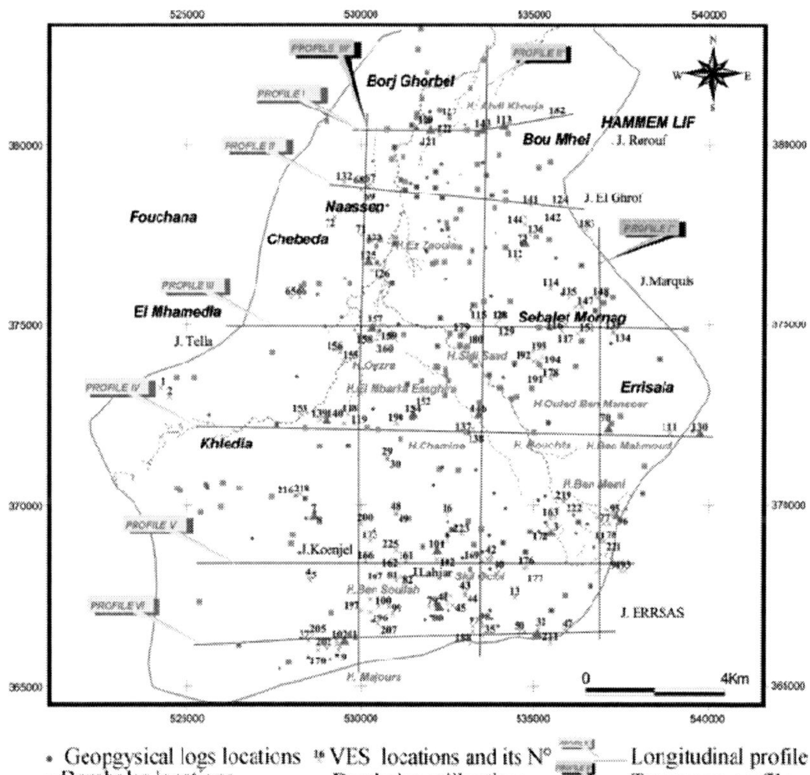

Figure 5. Location of vertical electrical soundings (VES) and the geoelectric cross-section (profiles).

It may include two superficial geoelectrical levels with variations in the thickness. The second layer corresponding to conductive level (C1) with a low true resistivity (3-13 Ω.m) and thickness ranging from 1 to 15 m. Low resistivity values of this layer can be attributed to the fine-grained lithologies, as indicated from the borehole data, consisting mainly of clay mixed with some sands. In the southern part of the area, this layer is occasionally absent. The third layer attains a wide true resistivity range, which varies from 35 to 148 Ω.m and the thickness ranges from 17 to 50 m.

This layer consists of medium to coarser grained lithologies which constitutes the first aquifer (R1). The fourth layer is conductive (C2) and shows a limited range of resistivity (6 to 10 Ωm) and thicknesses (8 to 18 m). This range of resistivity suggests a homogeneous lithology composed mainly of clay. This layer is occasionally absent, especially in the South, giving rise to a relatively higher resistivities sediments. The fifth layer is the lower most interpreted zone, which extends from the base of the overlying layer to the maximum depth of investigation.

Thus, the thickness of this layer is not defined. This bottom layer attains a wide resistivity range, which varies from 40 to 108 Ω.m. The interpretation of this resistivity range, in the light of correlation with the borehole information and field observations, indicates a heterogeneous lithology composed mainly of gravels, conglomerate and sands of variable grain size, and some clay. This layer corresponds to the main aquifer (R2) in the Mio-Plio-Quaternary series.

4.2.1. Transverse Geoelectrical Cross-Section along the Profile III

The correlation made on this profile (Figure 6a) made it possible to highlight two electrical discontinuities (F1 and F2), forming structural graben in the central part, which led to strong lateral and vertical variations in thickness and depth of the lithostratigraphic units.

The results of the VES are discussed utilizing a number of geoelectric profiles (Figure 6) and iso contour maps (Figures 7 and 8) to display the variations inherent (vertically and laterally) in the electrical resistivities.

The geoelectrical sections discriminate up the geoelectrical layers of different true resistivity ranges, thicknesses, depths, and hydrogeological characteristics. (Figure 6).

Nine geoelectrical cross sections were established for two main directions; W-E and N-S (Figure 5) Data from boreholes were integrated in these sections. In the western corner of the study area, four subsurface layers were recognized. The surface layer composed by the alternation of highly resistive and relatively low resistive lithologies (Figure 6a).

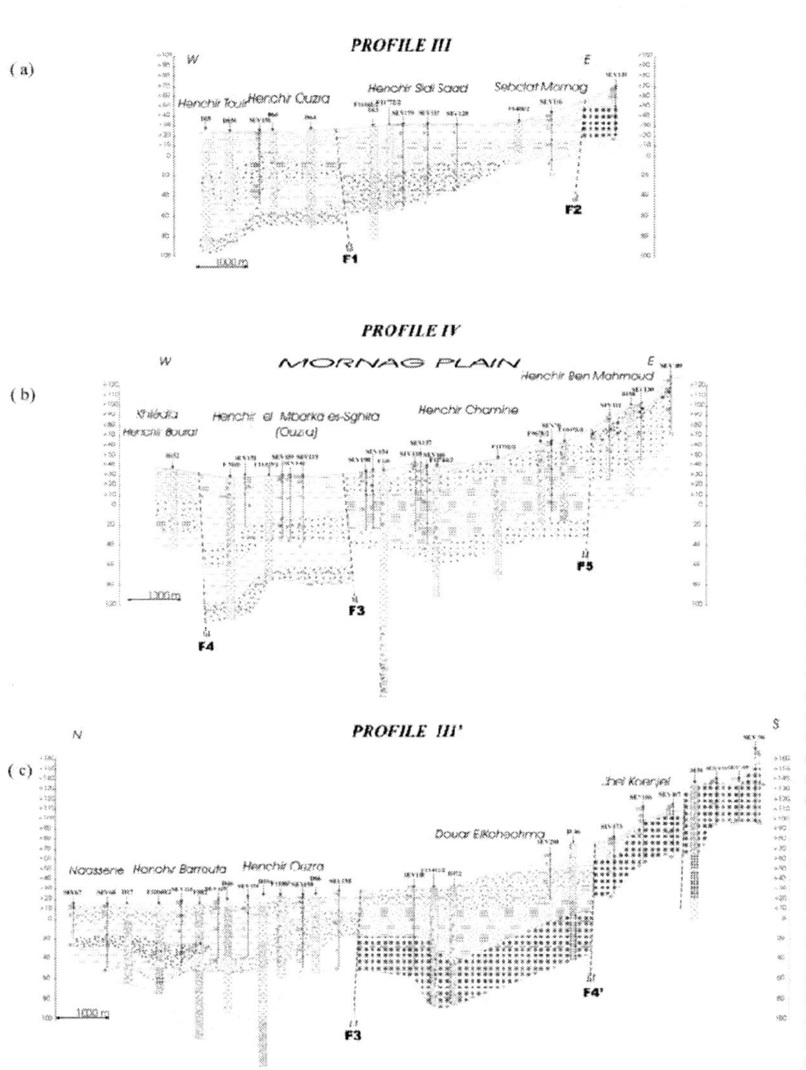

Figure 6. Iso- apparent resistivity contour maps for AB=2 of 50, 100, 200, 300 and 400 m in Mornag plain.

It ranges from 30 to 40 m in thickness and is composed of sand and clays intercalations. The second geoelectric layer corresponds to fine grained sand with small clay content that can constitute the main aquifer (R2). The average depth of this aquifer is between 46-80 m. The third geoelectric layer extends to a depth of about 90 m, with resistivity values ranging from 4 to 12 Ω.m and composed of a clay and clayey-sandy sequence (C2). The bottom layer corresponds to coarse grained sand that can constitute the semi-deep aquifer (R3).

In the central part, the structural graben of Mornag delimited by the faults F1 and F2 explain the thickening of the Mio-Plio-Quaternary series in this trough. The structural pattern influences not only the variations of the aquifer's depth and thickness, but has also a direct effect on the hydrogeological characteristics (Figure 6a). However, we note the absence of the conductive level (C2) giving rise to the second aquifer (R2) resting directly on the semi-deep aquifer (R3). This is justified by the well developed communication between these aquifers.

In the eastern corner of the study area, tow layers of different resistivity were detected: The first one is characterized by a wide resistivity range (7 to 120 Ω.m), reflecting a very heterogeneous lithology composed mainly of gravels and sands of heterogeneous grain sizes, including limestone fragments and some clay. The second layer, is characterized by high true resistivities (130 Ω.m) corresponding to coarse sand, conglomerate and gravel.

It can be deduced that the western part is relatively lower than the eastern part. The recorded erratic distribution in its thickness and depths clearly reflects the role of faults in controlling these variations. It is commonly assumed that the thickening of the aquifer series is influenced by the fault F1 and the downfall of the Mio-Plio-Quaternary series in the central part of the study area is controlled by the fault F2.

4.2.2. Transverse Geoelectrical Cross-Section along the Profile IV

This profile delineates a major faults (F3, F4 and F5) pattern, forming structural graben in the central part of the area, in which the maximum depths (from 130 to <150m) are recorded. In the remaining area, the aquifers becomes generally shallow with depths of <60m below the soil surface.

At the western corner of the profile (Figure 6b), two layers with dominating clay composition are observed in the borehole D152. The third layer, situated in intermediate position, is about 37 m thick and consists of gravel and medium to coarse grained sand. It constitutes the main aquifer.

Between F4 and F3: Below the superficial level, we find a relatively resistant layer (21-30 Ω.m) that could present a hydrogeological interest. It correspond to a sand and sandy-clayey sequence. The third geoelectrical layer is characterized by low resistivity (4-12 Ωm) reflecting the fine-grained compositions that should consist of clays, as indicated by the borehole data.

The thickness varies from 50 to 35m. The fourth geoelectric layer is characterized by a limited range of resistivities (67 to 51 Ωm) and thicknesses (18 to 25m). Its composed of medium to coarse grained sand that can constitute the main aquifer (R2). The true electrical resistivity map of the fifth geoelectric layer, shows a similar pattern of lithologic composition to that of the third geoelectric level, but with relatively low and limited values of resistivities (8 to 11 Ωm), which increase to the East. A bottom layer has not been reached by the soundings, because its depth may be greater than the maximum depth of investigation. Thus, in the light of correlation with the borehole information, it indicates a heterogeneous lithology composed mainly of sand that could make of the good sandy semi-deeper aquifer (R3).

To the East of the fault F3, five layers were recognized. The first conductive superficial level is fairly heterogeneous. This layer thins towards the east until its disappear (at F111702/2) giving rise to a resistant layer corresponding to coarse-grained lithologies. The second geoelectric layer is characterized by high true resistivities (40 to 60 Ω.m) and a thickness ranging from 17 to 25 m. The resistivity is diagnostic of coarse grained sand which constitutes the first aquifer (R1). The third geoelectric layer is thick and relatively resistant (21-27Ω.m).

It is composed of sandy and sandy-clayey sequence that could present a hydrogeological interest. The fourth conductive level is intercalated within the main aquifer subdividing it in two sub-units. Hydrologically, its position between the two resistive layers would certainly decrease the vertical relations by drainage between these two sub-units of the main aquifer (R2). The fifth geoelectric layer is the lowermost interpreted layer, which extends from the base of the overlying layer to the maximum depth of investigation. According to borehole information, this layer is composed mainly of sand and gravel that could correspond to the semi-deep aquifer (R3).

The thickness of this aquifer abruptly decreases towards the western part. In the eastern corner of the profile: Below a superficial highly resistive level (148 to 192 Ω.m), two layers that composed of medium to coarse grained sand are encountered: the first, is a relatively resistant level (50-72 Ω.m) with a maximum thickness not exceeding 20 m, The second is thicker (45m) and characterized by lower resistivity values (35-29 Ω.m).

Interpretation of these values, in the light of correlation with the borehole data, indicates that the lithology corresponds to a sandy-sandstone sequence which constitutes an aquifer of very good quality groundwater. The bottom layer, is characterized by low resistivity values (9 to 11 Ω.m), corresponding to fine-grained lithologies. The thickness of this substratum ranges from 35 to 40 m.

It is important to note that the abrupt changes in the thickness and depths of the interpreted geoelectric layers clearly indicate the presence of a number of faults affecting most of the sequence. This explains the presence of the same geoelectric layer at different depths, thickness variations and variable distribution of the layers all over the area.

This cross-section allowed us to follow faults that reflect different hydrogeological conditions between eastern and western parts and play an important role in the hydrodynamic behaviour of the aquifers, which is justified by the well developed communication (vertical and horizontal) between the three aquifers, especially in the part limited by F3 and F5.

4.2.2. Longitudinal Geoelectrical Cross-Section along the Profile PIII'

This profile exhibits some particular characteristics when compared to the other profiles. The faults F3 and F4, identified along this profile (Figure 6c), make it possible to consider the existence of blocks system. Indeed, the Mio-Plio-Quaternary aquifer is compartmentalized in to three collapsed blocks, which would explain the aquifers thickening from south to the north.

In the southern block, the first layer is characterized by the succession of several thin levels with contrasted resistivities. The second geoelectric layer extends from the base of the first layer to a depth ranging from 50 to 72 m. Its thickness ranges from <70 m, on the uplifted areas, to <35m in the collapsed areas (Figure 6c). It has higher resistivities ranging from 130 to 204 Ω.m, corresponding to Oligocene coarse sand and sandstones that can constitute a good hydrogeological target.

4.3. Thickness and Depth Contour Maps

The results of the VES method combining geophysical and lithological logs are discussed utilizing a number of iso-contour maps (Figures 7 and 8).

These maps illustrate lateral variations in depths and/or thicknesses of the interpreted geoelectrical layers that can constitute good hydrogeological objective (R1, R2 and R3) in the Mio-Plio-Qauternaire aquifers of the study area. Examination of these maps shows a number of faults that control the strong lateral variations in thickness and depth of the lithological units.

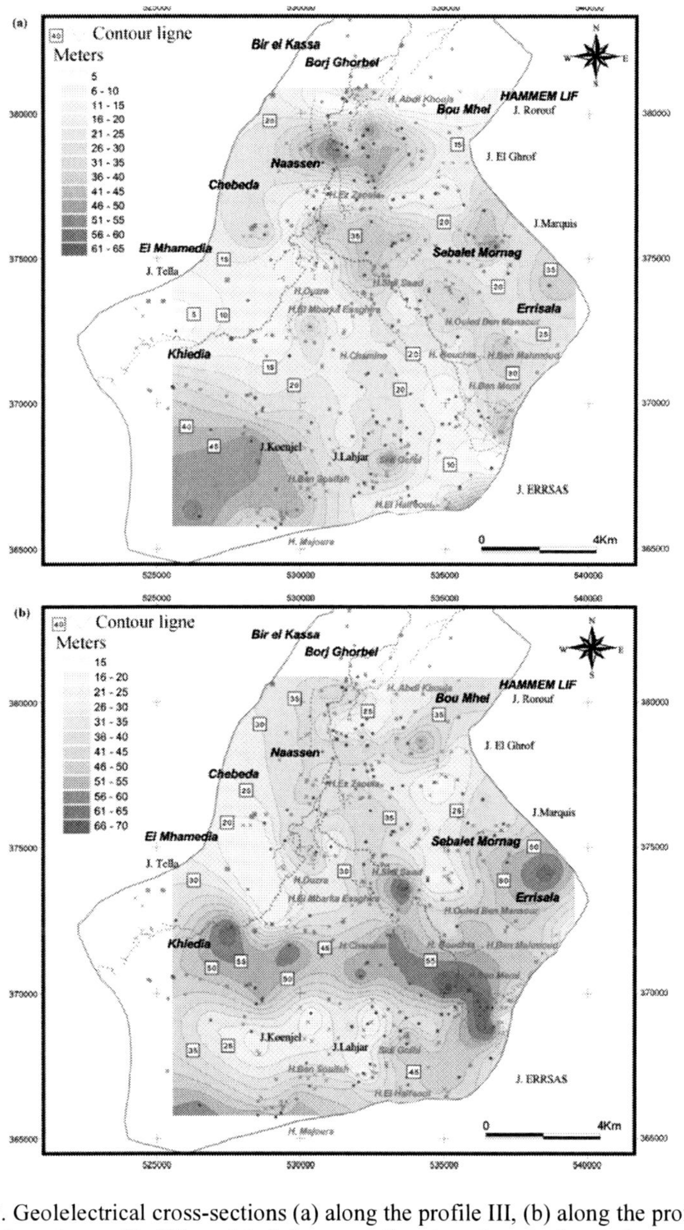

Figure 7. Geolelectrical cross-sections (a) along the profile III, (b) along the profile IV, and (c) along the profile III', and their interpretations based on the borehole calibrations.

4.3.1. Isopach Contour Maps of the Shallow Aquifer (R1)

The shallow level, supplied by the erosion of the surrounding reliefs (Cretaceous, Eocene et Oligo-Miocene), is characterized by a heterogeneous lithology composed mainly of sands of different grain sizes, interbedded with gravely sand, conglomerate, sandy clay and some clay. It constitutes the reservoir of the shallow aquifer. The isopach contour maps of this aquifer level show a wide range of thickness, which varies from 5 to 65m, with increasing values towards the East, South-West and North, and decreasing values generally towards the North-East and the central part of the plain, especially on both sides of the Miliane and El Hma rivers (Figure 8a).

Due to the shallow depth of this aquifer (10 to 55m), especially in the north-western and the central parts of the study area, a continuous rise of the groundwater table occurs.

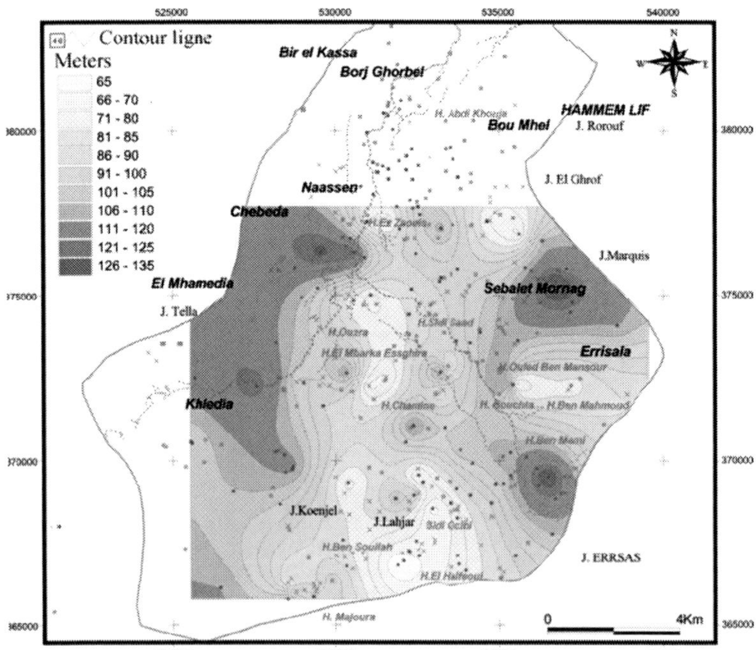

Figure 8. (a) Thickness contour map of the shallow aquifer (R1) based on interpretation of geophysical logs and VESs. (b) Thickness contour map of the main aquifer (R2) based on interpretation of geophysical logs and VESs.

4.3.2. Isopach Contour Maps of the Main Aquifer (R2)

The layer R2 could, locally, contain a good aquifer whose lithological nature could correspond mainly to a sandy sequence. This aquifer appears to be of erratic distribution in thickness. However, the isopach contour maps of show a minimum thickness (25m) of this aquifer around Sebalet Mornag and Chebeda localities and at the northern border of the Oligo-Miocene series cropping out at Khledia (Figure 7).

The aquifer thickness abruptly increases (up to 70 m) in the eastern and western parts of the study area, around Errisala and Khledia localities, and in the upstream of the Miliane and El Hma rivers. This pattern of distribution delineates subsurface fault patterns. The faults gave rise to preservation of a large thickness of this layer and the collapse of the clayey substratum in graben-like structures. Due to these subsurface complexities in the distribution of this layer and due to its hydrogeological importance as the main shallow groundwater aquifer in the area, care has to be taken in the site-selection for drilling productive wells.

4.3.3. Depth Contour Map of the Semi-Deep Aquifer

This level is the lowermost interpreted layer, which extends from the base of the overlying clayey zone to the maximum depth of investigation. Its lithology is described as sand and sandstone interbedded with sandy clay. The depth contour map of the top of this aquifer shows a wide range of depths, varying from 65 to 135 m. The maximum depths of this level (from 110 to <135m) are recorded in the Western, Southeastern and Northeastern parts of the area (Figure 8).

In the remaining area, the top of this aquifer becomes generally shallow with depths of <70m below the soil surface. Based on the interpreted lithological nature of this layer, in addition to its greater depths all over the study area, it may correspond to a relatively semi-deep aquifer which constitutes a good hydrogeological objective. The recorded erratic distribution in its thickness and depths clearly reflects the role of subsurface faults in controlling these variations.

4.4. Structural Map

Examination of the geophysical and geoelectrical cross sections, depth and thickness maps of the aquifer layers helps to interpret the subsurface structure of the Mio- Plio-Quaternary series.

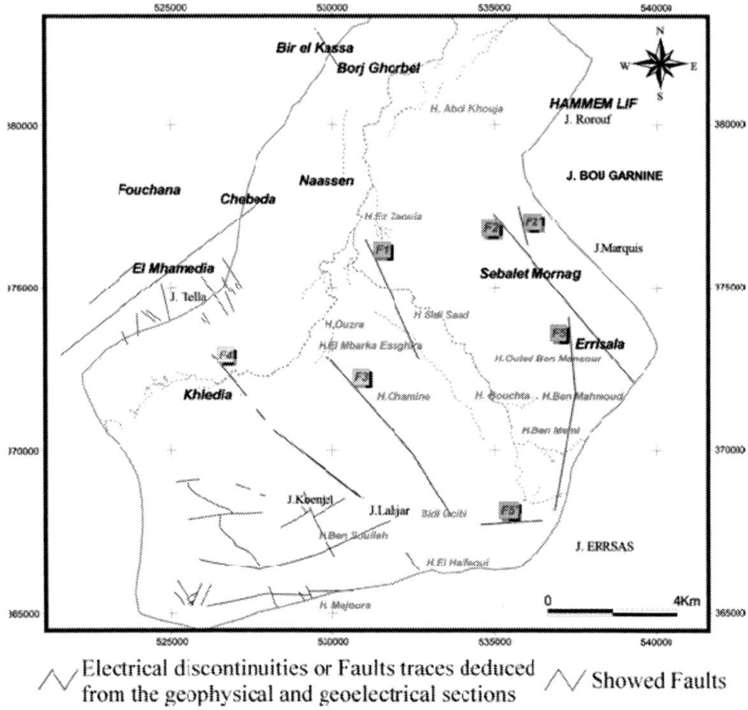

Figure 9. Depth contour map of the semi-deep aquifer, based on interpretation of geophysical logs and VESs.

The abrupt changes in the thickness and depths of the aquifer layers under consideration clearly indicate the presence of a number of faults affecting most of the sequence. This explains the presence of the same layer at different depths, thickness variations in the levels, and variable distribution of the zones all over the area. The maximum thicknesses are recorded in the graben structures or on the downthrown sides of the faults, while the minimum thicknesses are recorded on the upthrown sides of the faults (Figures 4 and 6). During this work several faults, in the NW-SE and NS directions, were identified (Figure 9).

SUMMARY

The combined use of geophysical logs and VES has provided hydrogeological information on the Mornag plain. These methods were successfully used to evaluate the subsurface hydrogeological and structural settings and to determine the aquifer geometry. However, interpretation of these methods indicates the presence of three hydrogeological facies corresponding to the shallow, the main and the semi-deep aquifers. In addition, subsurface structural complexities (faulting) were detected, which influenced the entire sedimentary section in the study area leading to strong lateral and vertical variations in the distribution of the subsurface lithological units. The geophysical information presented in this paper can be therefore considered as a basic documentation for future advanced hydrogeological studies in Mornag plain. The geoelectrical approach presented was successfully applied, and can be therefore applied in similar areas.

The structural pattern influences not only the variations of the aquifer's depth and thickness, but has also a direct effect on the hydrological characteristics, especially in the central part of the area.

References

Ahmed S., Marsily G., and Talbot A., (1988), Combined use of hydraulic and electrical properties of an aquifer in a geostatistical estimation of transmissivity. *Groundwater*, 26 1, 78-86.

Albouy Y., Andrieux P., Rakotondrasoa G., Ritz M., Descloitres M., Join J.L., Rasolomanana E., (2001), Mapping coastal aquifers by joint inversion of DC and TEM soundings-three case histories. *Groundwater*, 39, 87–89.

Asfahani J., (2007), Geoelectrical investigation for characterizing the hydrogeological conditions in semi-arid region in Khanasser valley, Syria. *Journal of Arid Environment*, 68, 31-52.

Astier J.L., (1971), Géophysique appliquée à l'hydrogéologie, Masson, Paris, 272 p.

Ayvaz M.T., (2009), Application of Harmony Search algorithm to the solution of groundwater management models. *Advances in Water Resources*, 32, 916-924.

Bowling J.C., Harry D.L., Rodriguez A.B., Zheng C., (2007), Integrated geophysical and geological investigation of a heterogeneous fluvial aquifer in Columbus Mississippi. j *Applied Geophysics*, 62, 58-73.

Bradford J.H., Sawyer D.S., Zelt C.A., Oldow J.S., (1998), Imaging a shallow aquifer in temperate glacial sediments using seismic reflection profiling with DMO processing. *Geophysics*, 63, 1248-1256.

Cardimona S., Clement W.P., Kadinskey C.K., (1998), Seismic reflection and ground penetrating radar imaging of a shallow aquifer. *Geophysics*, 63, 1310-1317.

Crumbling D.M., Griffith J., Powell D.M., (2003), Improving decision quality: making the case for adopting next generation site characterization practices. Remediation. *The Journal of Environmental Cleanup Costs, Technologies and Techniques*, 13(2), 91-111.

Custodio E., Bruggemen, G.A., and Cotecchia, V., (1987), Groundwater problems in coastal areas, Studies and Reports in Hydrology, 35, UNESCO, Paris.

Demlie M., Ayenew T., Wohnlich S., (2007), Comprehensive hydrological and hydrogeological study of topographically closed lakes in highland Ethiopia: The case of Hayq and Ardibo. *Journal of Hydrology*, 339, 145-158.

Dobrin, M.B., (1976), Introduction to Geophysical Prospecting. McGraw-Hill, New York.

Ebraheem A.A., Senosy M.M., Dahab K.A., (1997), Geoelectrical and hydrogeochemical studies for delineating ground-water contamination due to salt water intrusion in the northern part of the Nile delta, Egypt. *Groundwater*, 35, 216-222.

Ennabli M., (1977), Etude sur modèle mathématique des aquifères du Nord-Est de la Tunisie. Ecole. Nationale Supérieure des Mines, Paris.

Ennabli, M., (1980), Etude hydrogéologique des aquifères du Nord-Est de la Tunisie pour une gestion intégrée des ressources en eau. PhD thesis, Univ. Paris.

Essahlaoui A., Sahbi H., Bahi L., ElYamine N., (2001), Reconnaissance de la structure géologique du bassin de Saiis occidental, Maroc, par sondages électriques, *Journal of African Earth Sciences*, 32 (4), 777-789.

Horriche F., (2004), Contribution à l'analyse et à la rationalisation des réseaux piézométriques., PhD thesis, Université Tunis El Manar, ENIT, Tunisie, 254 p.

Giustiniani M., Accaino F., Picotti S., Tinivella U., (2009), 3D seismic data for shallow aquifers characterisation. *Journal of Applied Geophysics*, 68 (3), 394-403.

Grassi S., Cortecci G., (2005), Hydrogeology and geochemistry of the multilayered confined aquifer of the Pisa plain (Tuscany central Italy). *Applied Geochemistry*, 20, 41-54.

Gouasmia M., Gasmi M., Mhamdi A., Bouri S., Ben Dhia H., (2006), Prospection géoélectrique pour l'étude de l'aquifère thermal des calcaires récifaux, Hmeïma–Boujabeur (Centre-Ouest de la Tunisie), *Comptes Rendus Géoscience*, 338, 23-35.

Goldman M., Rabinovich B., Gilad D., Gev I., Schirov M., (1994), Application of the integrated NMR–TDEM method in the groundwater exploration in Israel. *Journal of Applied Geophysics*, 31, 27-52.

Harry D.L., Koster J., Bowling J.C., Rodriguez A.B., (2005), Multichannel analysis of surface waves generated during high resolution seismic reflection profiling of a fluvial aquifer. *Journal of Environmental and Engineering Geophysics*, 10 (2), 123-134.

Heigold P.C., Gilkeson R.H., Cartwright K., Reed P.C., (1979), Aquifer transmissivity from surficial electrical methods. *Groundwater*, 17 (4), 338-345.

Jones J.A.A., (1999), Climate change and sustainable water resources: placing the threat of global warming in perspective. *Hydrological Sciences Journal*, 44, 541-557.

Kelly W.E., (1977), Geoelectric sounding for estimating aquifer hydraulic conductivity. *GroundWater*., 15 (6), 420-425.

Kelly W.E., Frohlich R.K., (1985), Relations between aquifer electrical and hydraulic properties. *Groundwater*, 23 (2), 182-189.

Krivochieva S., Chouteau M., (2003), Integrating TDEM and MT methods for characterization and delineation of the Santa Catarina aquifer (Chalco Sub-Basin, Mexico). *Journal of Applied Geophysics*, 52, 23-43.

Kelly W.E., Maresˇ S., (1993), Developments in Water Science 44: Applied Geophysics in Hydrogeological and Engineering Practice, London, p. 289.

Kim H.S., Kim J.Y., Kim Y.S., (2004), The introduction of successive CMP refraction staking method and its application to investigate groundwater table in alluvial aquifer. Proceedings for 2004 Spring Meeting of Korea Society of Soil and Groundwater Environment (in Korean).

Kim H.S., Kim J.Y., (2008), High-resolution profiling of alluvial aquifer in potential riverbank filtration site by use of combining CMP refraction and reflection seismic methods. *Journal of Applied Geophysics*, 66, 1-14.

Koefoed O., (1979), Geosounding principle. Resistivity souding measurements,tome 1, Methods in Geochemistry and Geophysics, 14 A, p. 277.

Krivochieva S., Chouteau M., (2003), Integrating TDEM and MT methods for characterization and delineation of the Santa Catarina aquifer (Chalco Sub-Basin, Mexico). *Journal of Applied Geophysics*, 52, 23- 43.

Ladjmi T., (1960), Contribution à l'étude géologique et hydrogéologique de la plaine de Mornag. Mémoire DES. Fac. Sc. Uni. Paris, pp.68.

Lassoued B., Labbene H., Gouider N., (1995), Note sur l'exploitation du système aquifère de Mornag. Tunis. DGRE, pp.11.

Legchenko A., Baltassat J.M., Bobachev A., Martin C., Robain H., Vouillamoz J.M., (2004), Magnetic resonance soundings applied to characterization of aquifers. *Groundwater*, 42 (3), 436-473.

Legchenko A., Baltassat J.M., Beauce A., Bernard J., (2002), Nuclear magnetic resonance as a geophysical tool for hydrogeologists. *Journal of Applied Geophysics*, 50, 21-46.

Lubczynski M., Roy J., (2004), Magnetic Resonance Sounding: new method for ground water assessment. *Groundwater*, 42, 291-303.

Mazac O., Císlerová M., Vogel., T (1988), Application of geophysical methods in describing spatial variability of saturated hydraulic conductivity in the zone of aeration. *Journal of Hydrology*, 103 (1–2), 117-126.

McNeill J.D., (1980), Electrical conductivity of soil and rocks. Technical Note TN-5, Geonics Limited, Mississauga, Ont., Canada, p. 22.

Parasnis D.S., (1997), Principles of Applied Geophysics, Chapman and Hall, London, p.429.

Perry, E., A. Paytan, B. Pedersen, and G. Velazquez Oliman (2009), Groundwater geochemistry of the Yucatan Peninsula, Mexico: Constraints on stratigraphy and hydrogeology. *Journal of Hydrology*, 367, 27-40.

Pham V.N., Boyer D., LeMouël J.L., Thoa Nguyen T.K., (2002), Hydrogeological investigation in the Mekong Delta around Ho-Chi- Minh City (South Vietnam) by electric tomography, *Comptes Rendus Geoscience*, 334, 733-740.

RaidlaV., Kirsimäe K., Vaikmäe R., Jõeleht A. Karro E, Marandi A., Savitskaja L., (2009), Geochemical evolution of groundwater in the Cambrian–Vendian aquifer system of the Baltic Basin. *Chemical Geology*, 258, 219-231.

Ritzi R.W., Andolsek R.H., (1992), Relation between anisotropic transmissivity and azimuthal resistivity surveys in shallow, fractured, carbonate flow systems. *GroundWater*, 30, 774-780.

Rubin Y., Hubbard S.S., (2005), Introduction to hydrogeophysics. In: Rubin, Y., Hubbard, S.S. (Eds.), *Hydrogeophysics*, 5, 3-21.

Savadogo M.A., Nakolendousse S., Diallo S., (1997), Etude comparable de l'apport des méthodes électromagnetique MaxMin et électriques dans l'implantation des forages gros débits dans les régions de socle cristallin du Burkina Faso. *Journal of African Earth Science*, 24 (1/2), 169-181.

Slater L.D., Sandberg S.K., Jankowski M., (1998), Survey design procedures and data processing techniques applied to the EM azimuthal resistivity method. *Journal of Environmental and Engineering Geophysics*, 3, 167-177.

Shaaban F.F. (2001), Vertical electrical soundings for groundwater investigation in northwestern Egypt: a case study in a coastal area. *Journal of African Earth Sciences*, 33, 673-686.

Schirov M.D., Legchenko A.V., Creer G., (1991), A new direct non-invasive groundwater detection technology for Australia. *Exploration Geophysics*, 22, 333-338.

Shtivelman V., Goldman M., (2000), Integration of shallow reflection seismics and time domain electromagnetics for detailed study of the coastal aquifer in the Nitzanim area of Israel. *Journal of Applied Geophysics*, 44 (2-3), 197-215.

Steeples D.W., Miller R.D., (1990), Seismic-reflection methods applied to engineering, environmental, and ground-water problems. In: Ward, S.H. (Ed.), Geotechnical and Environmental Geophysics 1: SEG. *Investigations in Geophysics*, 5, 1-30.

Sumanovac F., (2006), Mapping of thin sandy aquifers by using high resolution reflection seismics and 2-D electrical tomography. *Journal of Applied Geophysics*, 58, 144-157.

Tabbagh A., (2006), Geophysical prospecting for water resources in arid zones. *Comptes Rendus Geoscience*, 338, 1131-1132.

Vengosh A., Rosenthal E., (1994), Saline groundwater in Israel: its bearing on the water crisis in the country. *Journal of Hydrology*, 156(1-4): 389-430.

Vouillamoz J.M., Descloitres M., Bernard J., Fourcassié P., Romagny L., (2002), Application of integrated magnetic resonance sounding and resistivity methods for borehole implementation, a case study in Cambodia. *Journal of Applied Geophysics*, 50 (1-2), 67-81.

Vouillamoz M., Chatenoux B., Mathieu F., Baltassat J.M., Legchenko A., (2007), Efficiency of joint use of MRS and VES to characterize coastal aquifer in Myanmar. *Journal of Applied Geophysics*, 61, 142-154.

Vogelsang D., (1995), Environmental Geophysics: a Practical Guide. Springer-Verlag.

Wattanasen K. Elming S.Å., (2008), Direct and indirect methods for groundwater investigations: A case-study of MRS and VES in the southern part of Sweden. *Journal of Applied Geophysics*, 66, 104-117.

Yaramanci U., Lange G., Hertrich M., (2002), Aquifer characterisation using Surface NMR jointly with other geophysical techniques at the Nauen/Berlin test site. *Journal of Applied Geophysics*, 50, 47-65.

In: Aquifers: Types, Impacts and Conservation ISBN: 978-1-61942-091-5
Editors: Z. Ouakili, et.al, pp. 85-86

Chapter 4

GROUNDWATER INTENSIVE USE CASE STUDY: MANCHA ORIENTAL AQUIFER (SE SPAIN)

David Sanz, Juan José Gómez-Alday, Santiago Castaño, and Angel Moratalla
Remote Sensing and GIS group. Institute for Regional Development, University of Castilla - La Mancha, Albacete, Spain

Mancha Oriental Aquifer (MOA) is one of the most extensive carbonate aquifers largest (7,260 km^2) in Southern Europe. MOA is located in the southeast of Spain and is encompassed within the Jucar River Basin. Since the 80's, the exploitation of groundwater resource has become a key driver for the socioeconomic development of MOA. Groundwater resources have been used to supply urban and industrial uses, and irrigation agriculture which nowadays area currently exceeds 100,000 ha. Groundwater withdrawals runs at rate of more than 400 Mm^3/yr. The overexploitation has caused two major quantitative impacts, namely a steady drop in groundwater level and a reduction of MOA discharge to Jucar River. In these sense, MOA is presented as a shining example of the problems that the exploitation of groundwater for agricultural purposes is for the sustainable management of water resources and ecosystems associated surface.

The goal of this paper presents a methodology to hydrogeological and hydrogeochemical characterization of MOA. Specifically, a summary of the results of several publications on the MOA (Sanz et al. 2009 and 2011, Moratalla et al. 2009). Deep knowledge of hydrogeological system with intensive agriculture is a fundamental tool that is allowing the water authorities to quantify the effects of watershed management policies than with a view to sustainable use of groundwater resources.

REFERENCES

Moratalla A, Gómez-Alday JJ, De las Heras J, Sanz D, Castaño S. Nitrate in the Water-supply Wells in the Mancha Oriental Hydrogeological System (SE Spain). *Water Resour Manage* 2008;23:1621-1640.

Sanz D, Gómez-Alday JJ, Castaño S, Moratalla A., De las Heras J, Martínez Alfaro PM (2009). Hydrostratigraphic framework and hydrogeological behaviour of the Mancha Oriental System (SE Spain). *Hydrogeol J* 17:1375-1391.

Sanz D; Castaño S; Cassiraga E; Sahuquillo A; Gómez-Alday J; Peña S; Calera A (2011) Modeling aquifer-river interactions under the influence of groundwater abstractions in the Mancha Oriental System. *Hydrogeol J* 19:475-487

In: Aquifers: Types, Impacts and Conservation ISBN: 978-1-61942-091-5
Editors: Z. Ouakili, et.al, pp. 87-107

Chapter 5

STUDY OF VARIATION IN GROUNDWATER QUALITY IN ARID COASTAL AQUIFER IN SOUTH-EASTERN TUNISIA: USING MULTIVARIATE FACTOR ANALYSIS

Zouari Kamel and Trabelsi Rim
Laboratory of Radio-Analysis and Environment,
National School of Engineers of Sfax,
BP 1173. 3038 Sfax, Tunisia

ABSTRACT

The present chapter is focused on the Djeffara aquifer system located in the south eastern part of Tunisia. This aquifer system, shared between Tunisia and Libya, constitutes the main water resource in all the coastal plain. Its recharge, particularly in its northern part is maintained only thanks to large-scale groundwater inflow from the CI aquifer system which constitutes the largest confined aquifer in the northern Sahara sedimentary basin.

Under arid climatic conditions and an increasing water demand for irrigation, Djeffara aquifer system, presents different vulnerabilities to anthropogenic activities. Intensive exploitation of the aquifer during last years induced declining water levels, drying up springs and groundwater quality deterioration. In order to evaluate the factors that significantly influence groundwater quality, the multivariate statistical techniques were performed on 90 well representatives of groundwater samples gathered from Djeffara shallow aquifer. The collected samples were analyzed for physical-chemical parameters and stable isotope contents (^{18}O, ^{2}H). Principal component analysis (PCA) results identified three major principal components (PCs) representing 71% of cumulative variance. The contribution degrees of variation in a data set are 47% for 1st PC, 13% for 2nd PC, and 11% for 3rd PC of data set. The first factor represents salinization process with seawater intrusion. The second factor shows flushing of gypsum minerals from sedimentary rocks and nitrate contamination by irrigation return flow. The third factor revealed the mixing effect with deep groundwater of Continental Intercalaire aquifer occurring in the northern part of the basin.

These results suggest that the groundwater quality in the study area seems to be controlled by not only saltwater intrusion but by other effects such as irrigation return flows and mixing with old groundwater of the Continental Intercalaire aquifer.

Keywords: Groundwater quality, Principal component analysis (PCA), chemical parameters, stable isotope, saltwater intrusion, Djeffara aquifer, Continental Intercalaire, Tunisia

I. INTRODUCTION

Groundwater quality constitutes a very sensitive issue in arid and semi arid coastal areas that are densely populated since the lack of permanent surface water network due to dry climatic conditions. Both natural processes and human activities may have a significant impact on groundwater quality and further limit its use as water supply. Groundwater quality could be influenced by many factors including climate, aquifer lithology, surface water recharge, seawater intrusion (Pulido-Leboeuf, 2004), waste water pollution, etc. Therefore, geochemical and hydrochemical investigation of groundwater resources are important to observe the suitability of water for a particular use.

Different techniques have been used in attempt to evaluate water quality, essentially based on chemical ions correlation (Piper and Sholler diagram) (Freeze and Cherry, 1979) and some ionic ratio to predict the origin of the mineralization (Kim et al., 2003; Pulido-Leboeuf et al., 2003).

In recent years, with increasing number of chemical and physical variables of groundwater, a wide range of statistical methods are applied for proper analysis and data interpretation (Sanchez-Martos et al., 2001 ; Reghunath et al., 2002 ; Chen et al., 2007). The multicomponent statistical analysis is frequently used to characterize groundwater geochemistry and to compare the different chemical parameters of the water analyses and determine the relationships between them (Sânchez-Martos et al., 2001; Reghunath et al., 2002; Zeng and Rasmussen, 2005).

Principal component analysis (PCA) is a mathematical technique that is successfully applied to investigate problems in geological investigation, such as geochemistry (Chen et al., 2007), and environmental science (Liu et al., 2003). This statistical method has several advantages over classical graphical approaches as it takes into consideration a huge amount of spatial and temporal data of water quality.

The main advantage of this method is the reduction of variables dimension by providing the correlation among measured chemical variables and their multivariate patterns based upon the correlation or covariance matrix (Helena et al., 2000). So, it is principally used to group together different geochemical variables according to their degree of co-variation among samples. The geochemical interpretation of determined factors gives insights into the dominant processes, which may influence groundwater chemistry (Chen et al., 2007).

The objective of this study is to apply multicomponent statistical analysis to understand the factors affecting the groundwater quality of the Djeffara shallow aquifer in southeastern of Tunisia and to discriminate their influence area.

II. Study Area

Djeffara plain is located in the most populous and intensively agricultural region of Libya and southern Tunisia. The Tunisian part is bound in the north by the Skhira region, in the west by the El-Hamma faults, Matmatas and Dahar Mountains, and in the south by Libya boundaries. The Djeffara plain is largely open in the east on the Gulf of Gabes and the Mediterranean Sea (Figure 1).

The climate is classified as semi-arid to arid climate type. Average annual rainfall in the study area is 190 mm, which mainly takes place during the short cold season. The mean annual temperature is 20.6°C and the evapotranspiration rate exceeds 1700 mm/year. This situation of dryness provokes the loss of water resource especially during the last decade due to the fact that the renewal rate of this resource is very weak.

The Djeffara basin constitutes a NW-SE trending collapsed block which is strongly affected by tectonic deformations and faults in different directions indicating a complex structural history (Bouaziz et al., 2002; Gabtni et al.,

2009). El Hamma and Medenine faults are the most prominent features that bound the Djeffara aquifer from the west.

The geologic deposits surrounding Djeffara plain include sedimentary formation extending from the Palaeozoic one to the Quaternary. The thickness and the lithology of these formations vary widely from the north to the south and from the east to the west (Bouaziz, 1995; Bouaziz et al., 2002; Gabtni et al., 2009).

This lithological variation and the structural complexity of the Djeffara basin allowed the individualization of a multi-layer aquifer system placed in the carbonated formations of Senonian, Turonian and the Jurassic one, and the detrital formations of Mio-Plio-Quaternary.

The present study focuses on the shallow aquifer of Djeffara which is logged in Mio- Plio- Quaternary continental deposits extending over the coastal fringe.

The Quaternary deposits include an alternation of clay-sand layers and limestones passing to the detrital series of Mio-Pliocene. The total depth of exploiting wells varies from 3 m to 20 m where the transmissivity is about 5 x 10^{-5} m^2/s.

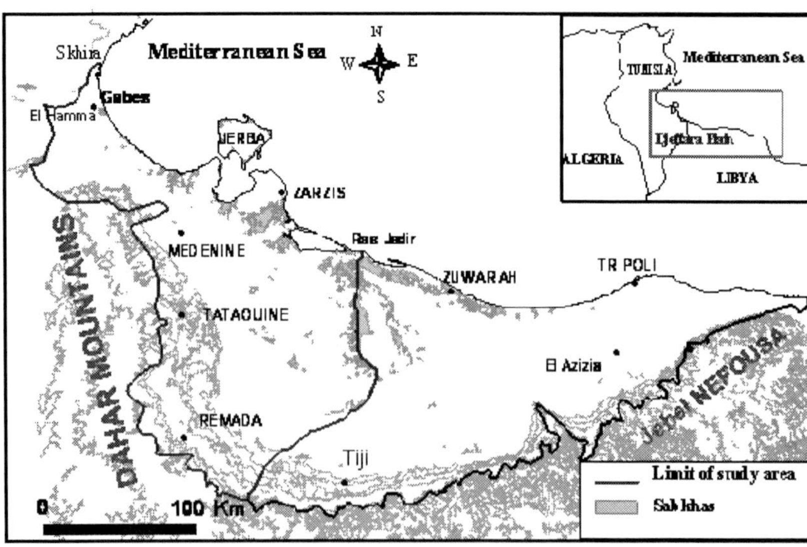

Figure 1. Location map of the Djeffara plain.

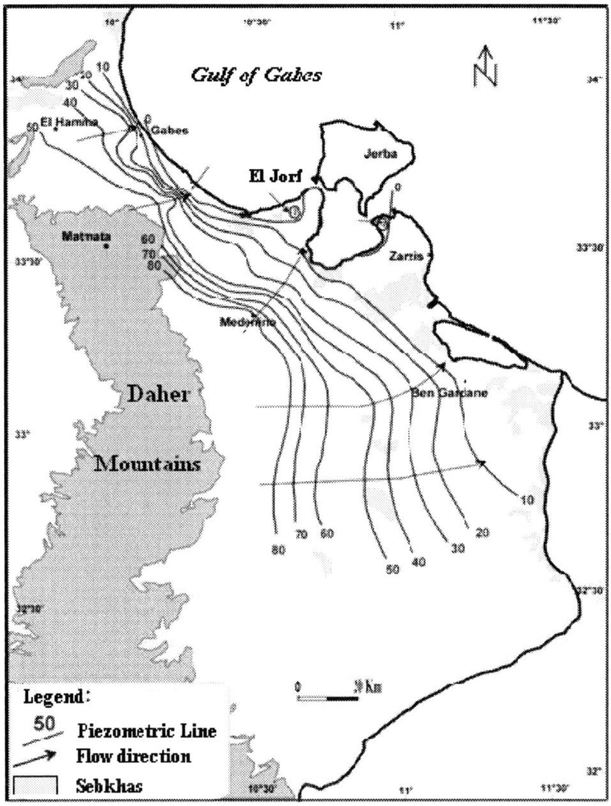

Figure 2. Piezometric map of Djeffara shallow aquifer in 2006.

A number of major hydrological and geochemical investigations (ERESS 1972; Gonfiantini et al. 1974; Mamou 1990; Trabelsi et al., 2009) have been carried out on this aquifer system. It has been established that the Continental Intercalaire, one of the largest confined aquifers in the world, is flowing by upward leakage to the Djeffara aquifer system through El Hamma fault in the north western part of the study area. These studies proved convincingly that the Continental Intercalaire represents a non-renewable groundwater resource, recharged during past pluvial periods which occured across entire Northern Africa during the late Pleistocene and early Holocene. However, the shallow coastal aquifer of the Djeffara plain is recharging primarily from direct infiltration of rainfall and excess irrigation water and partially by vertical leakage from the CI aquifer in El Hamma region. The piezometric map of the shallow aquifer of Djeffara (Figure 2) indicates a west-east flow direction from the Daher Mountains and El Hamma fault system located to the west, towards Mediterranean Sea, which constitutes the main discharge area.

Nevertheless, the significant socio-economic development of the study area, accompanied by a strong water demand, induced an increase of the exploitation of groundwater resources.

The Over-pumping of the shallow aquifer systems of Djeffara caused a general drawdown of the water table. That results in new hydrodynamic equilibrium characterized by migration of the seawater interface towards the coastal sectors and the increase in groundwater salinity especially close to the El Jorf and Jarzis areas (Figure 2).

The main objectives of the present study are (1) the evaluation of the chemistry and quality of groundwater (2) the determination of the natural and anthropogenic process that influences groundwater quality.

The integration of hydrological, geochemical and isotopic tracers with statistical methods is used for delineating hydrochemical processes, identifying the consequences of the water-rock interaction, the mixing with Continental Intercalcaire groundwater and Mediterranean seawater.

III. Materials and Methods

For the assessment of groundwater quality, 92 water samples were collected from the shallow aquifer of the Djeffara plain. Groundwater samples were analysed for chemical and isotopic compositions. Measurement of pH, Conductivity, TDS and alkalinity were measured in the field. Major elements (Cl^-, SO_4^{2-}, NO_3^-, HCO_3^-, Ca^{2+}, Mg^{2+}, Na^+, K^+) were analysed by ion liquid chromatography at the Laboratory of Radio-Analysis and Environment of the National School of Engineers of Sfax.

Stable isotope composition of the groundwater samples was determined by the liquid water analyser LGR based on off-axis integrated cavity output spectroscopy lasers (Lis et al., 2007) at the Laboratory of Radio-Analysis and Environment of Sfax and by the mass-spectrometry measurement Isotope Hydrology Laboratory of the International Atomic Energy Agency in Vienna, using CO_2 equilibration technique for $\delta^{18}O$ (Epstein and Mayeda, 1953) and isotopic exchange with H_2 gas for δ^2H (Colpen et al., 1991). All statistical evaluations were carried out using XL STAT 2006 for windows.

IV. RESULTS AND DISCUSSION

Hydrochemical Characteristics

Electrical conductivity (EC), T°, pH and major ion concentrations are summarized in table 1. The pH ranges from 6.6 to 8.5 with an average value of 7.6 which indicate that the groundwater is slightly alkaline. The temperature of groundwater samples varies widely, from 16°C to 45°C, independently of well depth. The water temperature is higher in the El Hamma region indicates some inflow of deep groundwater from the CI aquifer through existing faults. In fact, the groundwater samples belonging to the CI system are characterized by elevated temperatures, in the range between 50°C and 61.4°C Owing to the fact that CI is a confined aquifer system as the unsatured zone is thick and formed by clay- rich strata (Trabelsi et al., 2009). The electrical conductivity (EC) and total dissolved solids values range from 930 to 29526.8 μS/cm and 765.4 to 25708 mg/l with an average value of 3866.5 μS/cm and 3145.9 mg/l respectively (Table 1).

Table 1. Statistical summary of hydrochemical and isotopic parameters of groundwater samples

Variable	Maximum	Minimum	Average	Std.dev
pH	8.52	6.6	7.6	0.4
EC	29526.8	930	6319.7	3866.5
T°C	45.0	16	21.8	4.8
TDS (mg/l)	25708	765.4	5421.9	3145.9
Ca	1684.8	140.4	634.3	288.4
Mg	1451.6	8.8	171.7	174.4
K	318.9	0	36.3	56.1
Na	4381.3	78.4	904.8	655.2
HCO3	445.3	6.1	149.2	66.5
CO3	66.0	0	6.2	10.7
Cl	7597.4	129.3	1515.9	1213.1
NO3	427.3	0	17.9	54.5
SO4	7139.4	157.5	1826.5	918.8
$\delta^{18}O$	-1.7	-8.3	-5.3	1.5
$\delta^{2}H$	-12.3	-62.3	-35.8	11.2

pH (standard units), EC (μ.Siemens/cm), concentration (mg/l) and stable isotopes (‰ vs SMOW).

These significant variations in groundwater salinity could be related to different process such as evaporation, anthropogenic activity, water rock interaction and mixing process. The order of abundance of the cation chemistry for are Ca _ Na > Mg > K and all groundwater samples exceed the suitable limit of Ca for drinking water (75mg/l) (WHO, 1993). The order of anion chemistry is Cl _ SO_4 > HCO_3.

All samples groundwater samples exceed the suitable limit of Cl (200 mg/l) for drinking water (WHO, 1993), except for sample P90 where Cl content is 129.3 mg/l. In order to compare between different types of waters, groundwater samples are plotted on a Piper diagram (Figure3). The groundwater samples evolve between two types: (1) sodium chloride water type it regoups the most saline groundwater samples located particularly in coastal area and (2) sulphate –calcium water type which regoups the less mineralized groundwater samples. The enrichment in chlorides and sodium present in the groundwaters samples located near the coastal area (Eastern part of the basin) may be caused by seawater intrusion, halite dissolution and partly evaporation /or irrigation return flow.

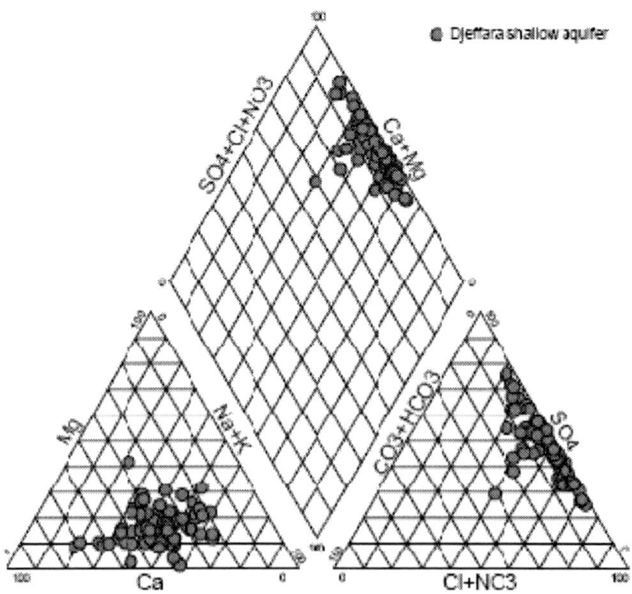

Figure 3. Piper diagram of the analysed groundwater samples.

Isotopic Data

The measured stable isotopes contents in groundwater samples range from -8.3 to -1.7 ‰ vs SMOW for $\delta^{18}O$ and -62.3 to – 12.3 ‰ vs SMOW for $\delta^{2}H$. Plotted in oxygen-18/ Deuterium diagram (Figure 4) together with seawater sample and the mean isotopic composition of the CI groundwater; Djeffara groundwater plots slightly below the the Global Meteoric Water Line (GMWL) (Craig, 1961) and the Regional Meteoric Water Line of Sfax (RMWL) (Trabelsi, 2009) which is located ca. 210 km north of the study area. The $\delta^{18}O$ and $\delta^{2}H$ compositions of freshwaters are close to the meteoric water line and plot along the RMWL.

These points represent groundwater collected in the Matmata reliefs and Daher mountains. Most likely, they represent recent infiltration water. The other samples points are distributed according to two mixing line:

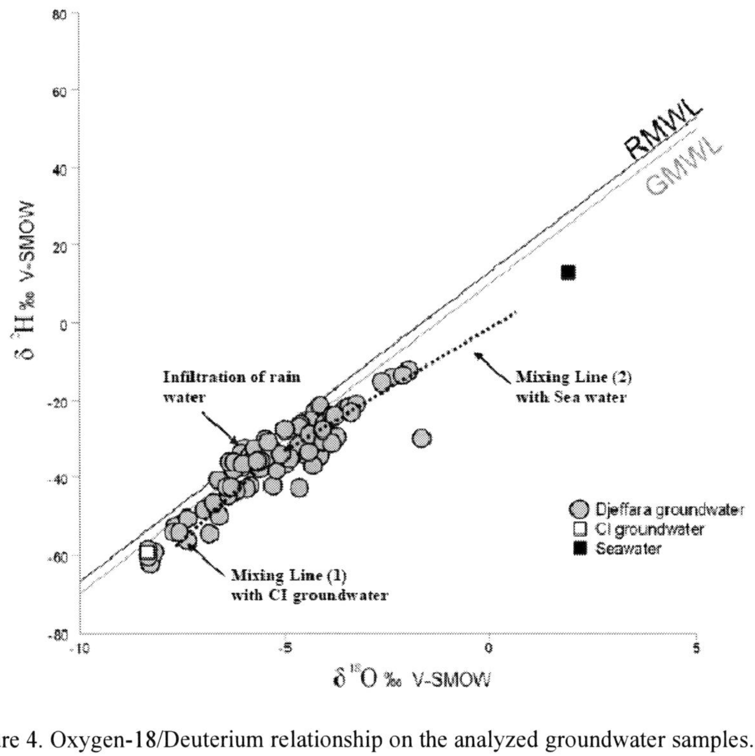

Figure 4. Oxygen-18/Deuterium relationship on the analyzed groundwater samples.

1) The first line represents a mixing with CI groundwater by upward leakage through the system of faults of El Hamma. The mean isotopic composition of CI aquifer is being -8.3 ‰ and -59.5‰, for $\delta^{18}O$ and $\delta^{2}H$ respectively (Figure 4). This depleted isotope signature is characteristic of old recharge occurring during the late Pleistocene period in northern Africa (Gonfiantini et al., 1974; Fontes et al., 1983; Zouari, 1988; Edmunds et al., 2003 and Zouari et al., 2003). In the El Hamma region of the Djeffara aquifer, the $\delta^{18}O$ and $\delta^{2}H$ values decrease gradually towards to the major fault where they become indistinguishable from the observed isotopic composition of the CI groundwater (Figure 4).
2) The second line represents a binary mixture between seawater and shallow aquifer. Most of samples plotting near this line are characterized by high chloride concentrations and an enriched stable isotopes composition compared with freshwaters. These points are located in the coastal area in particular in El Jorf and Zarzis regions.

V. Multivariate Statistical Analysis

Correlation Analysis

In order to quantitatively analyse and characterize the relationship between physico-chemical parameters and stable isotopes contents in groundwater samples, Pearson's correlation analysis was applied to the data (Table 2). EC and TDS Major display significant correlation with major elements such as Na, Cl, K, Mg and SO_4 (0.64<r>0.95). This reflects the fact that groundwater mineralization is strongly controlled, by these elements.

Significant positive correlations are observed between Na, Cl, K, Mg and SO_4 ($r > 0.5$). The strong correlation observed between Na and Cl ($r= 0.99$) indicate the same origin probably halite dissolution or salt water intrusion. Less significant correlation factor is observed between Ca and SO_4 ($r= 0.74$). Ca and Mg are not significantly correlated (0.29) indicating that Ca and Mg are not probably related to dissolution of low magnesium carbonate, or dolomite deposits.

Table 2. Pearson correlation among physicochemical properties and stable isotopes in Djeffara shallow aquifer

Variables	pH	EC	T°C	TDS	Ca	Mg	K	Na	HCO_3	CO_3	Cl	NO_3	SO_4	^{18}O	^{2}H
pH	**1.00**	0.08	0.00	0.06	-0.16	0.00	0.01	0.13	-0.08	0.34	0.11	0.20	-0.06	0.03	-0.04
EC	0.08	**1.00**	-0.05	**0.95**	0.29	**0.87**	**0.76**	**0.94**	0.00	0.27	**0.94**	0.36	**0.59**	**0.60**	**0.53**
T°C	0.00	-0.05	**1.00**	-0.10	-0.14	-0.14	-0.12	-0.06	-0.14	-0.02	-0.08	-0.09	-0.13	-0.38	-0.46
TDS (mg/l)	0.06	**0.95**	-0.10	**1.00**	0.44	**0.90**	**0.82**	**0.97**	0.03	0.23	**0.96**	0.36	**0.71**	**0.57**	0.49
Ca	-0.16	0.29	-0.14	0.44	**1.00**	0.29	0.27	0.30	-0.02	-0.29	0.31	-0.03	**0.74**	0.32	0.18
Mg	0.00	**0.87**	-0.14	**0.90**	0.29	**1.00**	**0.77**	**0.86**	0.15	0.22	**0.86**	0.33	**0.69**	**0.53**	**0.50**
K	0.01	**0.76**	-0.12	**0.82**	0.27	**0.77**	**1.00**	**0.85**	-0.01	0.37	**0.87**	0.31	0.47	**0.48**	0.45
Na	0.13	**0.94**	-0.06	**0.97**	0.30	**0.86**	**0.85**	**1.00**	0.03	0.31	**0.99**	0.37	**0.56**	**0.56**	**0.50**
HCO_3	-0.08	0.00	-0.14	0.03	-0.02	0.15	-0.01	0.03	**1.00**	-0.09	0.00	0.01	0.13	-0.08	-0.03
CO_3	0.34	0.27	-0.02	0.23	-0.29	0.22	0.37	0.31	-0.09	**1.00**	0.31	0.31	-0.11	0.17	0.18
Cl	0.11	**0.94**	-0.08	**0.96**	0.31	**0.86**	**0.87**	**0.99**	0.00	0.31	**1.00**	0.38	**0.53**	**0.58**	**0.51**
NO_3	0.20	0.36	-0.09	0.36	-0.03	0.33	0.31	0.37	0.01	0.31	0.38	**1.00**	0.02	0.28	0.29
SO_4	-0.06	**0.59**	-0.13	**0.71**	**0.74**	**0.69**	0.47	**0.56**	0.13	-0.11	**0.53**	0.02	**1.00**	0.33	0.23
^{18}O	0.03	**0.60**	-0.38	**0.57**	0.32	**0.53**	0.48	**0.56**	-0.08	0.17	**0.58**	0.28	0.33	**1.00**	**0.94**
^{2}H	-0.04	**0.53**	-0.46	0.49	0.18	**0.50**	0.45	**0.50**	-0.03	0.18	**0.51**	0.29	0.23	**0.94**	**1.00**

A correlation factor higher than 0.5 is considered significant.

Table 3. Results of Principal Component Analysis showing Eigenvectors of extracted components

	F1	F2	F3	F4	F5	F6	F7	F8	F9	F10	F11	F12	F13	F14	F15
eginvalue	7.055	1.958	1.575	1.088	0.914	0.718	0.586	0.457	0.293	0.189	0.070	0.050	0.037	0.008	0.002
% of Variance	47	13	11	7.254	6.092	4.787	3.907	3.045	1.956	1.258	0.465	0.334	0.247	0.052	0.011
% cumulative variance	47	60	71	77.85	83.94	88.72	92.63	95.68	97.63	98.89	99.36	99.69	99.94	99.99	100.00

Principal Component Analysis

Principal component analysis (PCA), is the most common multivariate statistical method used in geochemistry studies. It helps to reduce data and to extract a small number of latent factors for analyzing relationships among the observed variables.

The most common PCA type producing more interpretable components is the varimax rotation, which is applied in the current study. The total number of factors or components generated from a multivariate analysis indicates the total number of possible sources of variation in groundwater quality. Principal component axes were rotated to make high component loadings higher and low component loadings lower, enhancing interpretation of the data.

Factor loadings > 0.71 are typically regarded as excellent and < 0.32 very poor for interpretation (Garcia et al., 2004). Under this criterion, only factors with eginvalues greater than or equal to 1 will be selected as possible sources of variance in the data (Kaiser, 1960). The rationale for choosing 1 is that a factor must have a variance at least as large as that of a single standardized original variable to be acceptable.

The obtained factors of PCA of collected groundwater samples are ranged in order of merit (table 3). The first factor has the highest eigenvector sum and represents the most important source of variation in the data. The last factor is the least important process contributing to the chemical variation.

Three factors are extracted to statistically represent the contributions influencing chemical composition of groundwater samples. These components all describe 71% of the total variance of the data (table 3). It suggests that all hydrogeochemical process in the study area could not be completely explained by PC1, PC2 and PC3.

The factor loadings tables are interpreted as correlation coefficients between the variables and the factors.

- The first component (PC1), explaining 47% of the total variance, shows strongly and positively loading related to EC, TDS, Mg, K, Na, Cl, δ^2H and $\delta\ ^{18}O$ (table 4). The strong association of Na, Ca, Mg and Cl with EC on PC1 indicates that these ions have dominant contribution to the explanation of the PC1which characterize groundwater salinization

The presence of Na and Cl could be interpreted as a signature of the halite dissolution and/or saltwater intrusion. But the strong correlation with stable isotopes confirms the important influence of seawater intrusion on groundwater quality (Figure 5). Consequently, the movement of PC1 means the effect of saltwater intrusion

- The second principal component (PC2), explaining 13% of the total variance, has a high negative factor loading for Ca and SO4 and positive loading for CO_3, pH and NO_3. This factor does not show significant loading with EC and probably reflect anthropogenic inputs (Figure 5). It represents strong positive correlations that characteristic of agricultural pollution, soil weathering and subsequent runoff.
- The third component (PC3) explains 11% of the total variance (Table 4). It has high positive loading for T° and high negative loadings for ^{2}H and ^{18}O. This component is assumed to represent the mixing process with deep groundwater of CI aquifer characterized by depleted stable isotopes composition and elevated temperature (Figure 5).

In order to find the correlations between the sampling sites and the variables obtained by physical, chemical and isotopic measurements of Djeffara shallow aquifer, all groundwater samples are represented according to the three principal components (Figure 6).

The spatial distribution of the scores of factor 1 shows that high values are characterizing the southern coastal part of the Djeffara in particular Medenine region (Figure 6).

These values are related to the raise of groundwater salinity due to seawater intrusion process which causes an increase in Cl, Na, K and Mg contents. The low scores of factor 1 recorded around Matamata and Tataouine regions is related to the presence of fresh groundwater and confirm the contribution of the Dahar upland to the recharge of the Djeffara shallow aquifer. The second PC2 which represents anthropogenic pollution related to the agricultural activities, concerns groundwater samples collected from the southern part of the plain characterized by an important agricultural activities. These contamination reflected by the increase of NO_3 contents is favorised by the lithology of the unsatured zone and the small depth of shallow wells.

The movement of PC3 means the effect of mixing with CI deep groundwater. The high scores value of this principal factor are characterizing the northwestern part of the study area, close to the El Hamma fault system. It indicates the impact of the mixing process on groundwater quality.

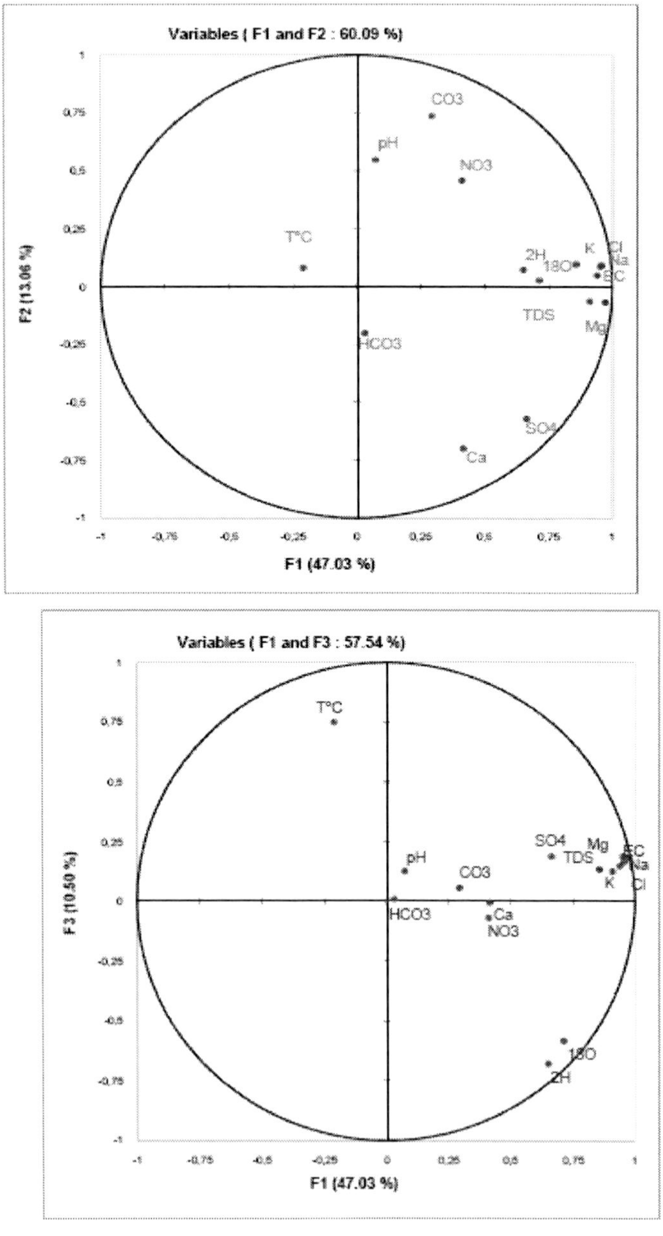

Figure 5. Scatter plot of loadings for F1, F2 and F3 factors obtained for all samples data set.

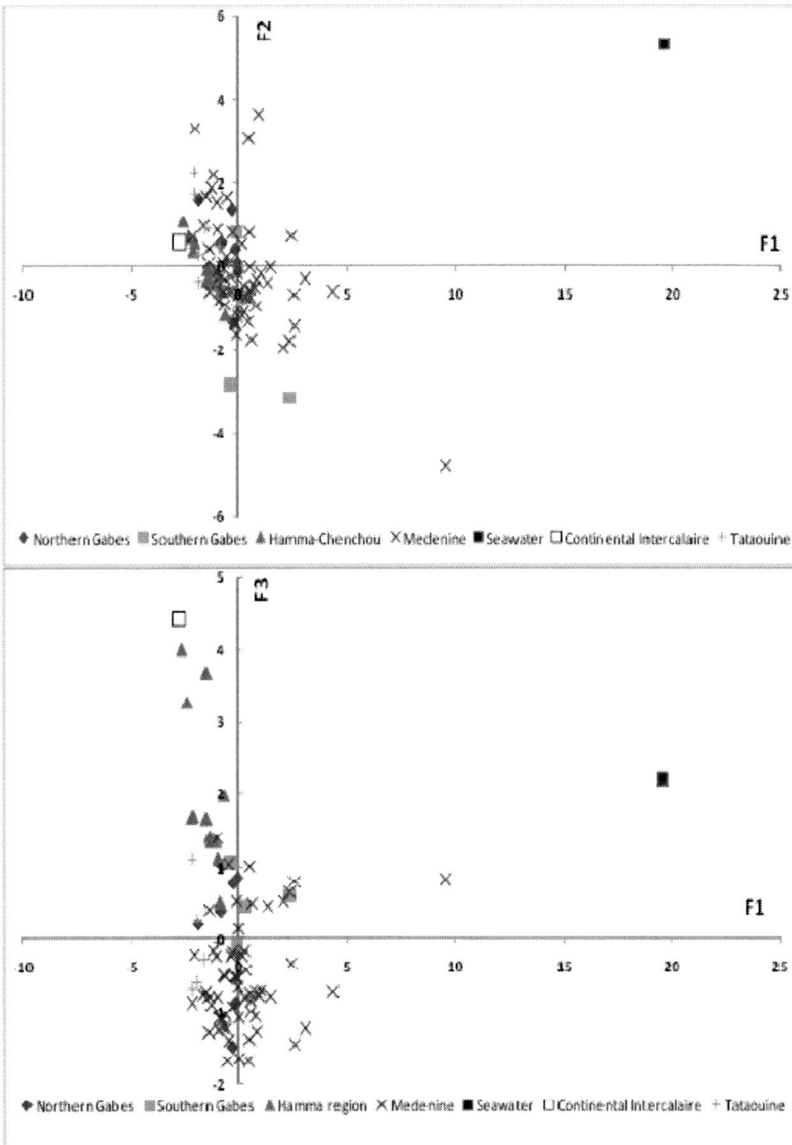

Figure 6. Scatter plot of scores for the three principal components obtained for all groundwater samples.

Table 4. Principal Component Analysis for groundwater samples

	F1	F2	F3
pH	0.073	**0.548**	0.126
EC	**0.940**	0.049	0.150
T°C	-0.211	0.082	**0.751**
TDS (mg/l)	**0.970**	-0.066	0.184
Ca	0.417	**-0.700**	-0.003
Mg	**0.909**	-0.063	0.127
K	**0.855**	0.097	0.136
Na	**0.953**	0.089	0.190
HCO_3	0.031	-0.201	0.010
CO_3	0.294	**0.736**	0.058
Cl	**0.956**	0.092	0.166
NO_3	0.412	**0.459**	-0.068
SO_4	**0.663**	**-0.571**	0.189
^{18}O	**0.712**	0.028	**-0.581**
^{2}H	**0.650**	0.074	**-0.678**

Rotation Method: Varimax with Kaiser Normalization.

CONCLUSION

This study has successfully demonstrated the utility of multivariate statistical tools including factor analysis to characterize the groundwater quality of Djeffara shallow aquifer.

The results obtained through this work help to find the correlations between the sampling sites and the variables obtained by physical and chemical measurements, which can be used to construct a fast decision model for separating different water quality samples.

The contribution degrees of variation of the 1st, 2nd, and 3rd PC are 47, 13, 11% of data set, respectively, and 71% of the total variance can be attributed to the three PC. The evolution according PC1 means impact by seawater intrusion which characterizes the southeastern part of the plain in particular in Medenine region. Also, the movement of PC2 means influence of agricultural activities reflected by high nitrate contents. In case of PC3, it reflects the mixing with CI deep groundwater occurring in the northern part of the Djeffara plain in particular in El Hamma region. These results suggest that the groundwater quality of the Djeffara shallow aquifer seems to be controlled by not only saltwater intrusion but by other effects such mixing with deep groundwater or irrigation return flow.

These informations could be helpful to official authorities in optimizing the groundwater management in southern Tunisia plan and enhancing their pollution control action.

REFERENCES

Bouaziz, S. (1995). Etude de la tectonique cassante dans la plate-forme et l'Atlas Sahariens (TunisieMéridionale): *Evolution des paléochamps de contraintes et implications géodynamiques.* PhD thesis. Fac. Sc. De Tunis. 301p.

Bouaziz, S., Barrier. E., Soussi, M., Turki, M.M., Zouari, H. (2002). Tectonic evolution of the northernAfrican margin in Tunisia from paleostress data and sedimentary record. *Tectonophysics* 357. 227–253.

Chen, K., Jiao, J. J., Huang, J., and Huang, R. (2007). Multivariate statistical evaluation of trace elements in groundwater in a coastal area in Shenzhen. China. *Environment Pollution.* Vol. 147. pp. 771-780. 2007.

Colpen, TB., Wildman, J., Chen, J. (1991). Improvement in the gaseous hydrogen-water equilibration technique for hydrogen isotopes ration analysis. *Anal. Chem.* 63, 910-912.

Craig, H. (1961). Isotopic variations in meteoric waters. *Science* 133, 1702–1703.

Edmunds, WM., Guendouz, A., Mamou, A., Moulla, A., Shand, P., Zouari, K. (2003) Groundwater evolution in the Continental Intercalaire aquifer of Southern Algeria and Tunisia: trace element and isotopic indicators. *Applied geochemistry* 18, 805- 822.

Epstein, S., and Mayeda, TK. (1953); Variations of ^{18}O content of waters from natural sources. *Geochim Cosmochim Acta* 4, 213-224.

ERESS (1972). Etude des ressources en eau du Sahara Septentrional. UNESCO. Paris.

Fontes, J.C., Coque, R., Dever, L., Filly, A., Mamou, A. (1983). Paléohydrologie isotopique de l'wadi el Akarit (Sud Tunsien) au Pléistocène at à l'Holocène. *Palaeolgeogr Palaeoclimatol Palaeoecol* 43, 41-61.

Freeze, R. A. and Cherry, J. A. (1979). *Groundwater.* Prentice-Hall. Englewood Cliffs. 604 pp.

Gabtni, H., Jallouli. C., Mickus, K.L., Zouari, H., Turki, M.M. (2009). Deep structure and cristal configuration of the Jeffara basin (Southern Tunisia) based on regional gravity. seismic reflection and borehole data: How to explain a gravity maximum within a large sedimentary basin?. *Journal of Geodynamics* 47 (2009) 142–152.

Garcia, J.H., W.W. Li, Arimoto, R., Okrasinski, R., Greenlee, J. and Walton, J. (2004). Characterization and implication of potential fugitive dust sources in the Paso del Norte region. *Sci. Total Environ*, 325: 95-112.

Gonfiantini, R., Conrad. G., Fontes, JC., Sauzy, G., Payne, BR. (1974). Etude isotopique de la nappe du Continental Intercalaire et ses relations avec les autres nappes du Sahara Septentrional. IAEASM- 182/25 In: *Isotope techniques in groundwater hydrology*. vol I. IAEA. Vienna. 227–241.

Helena, B., Pardo, R., Vega, M., Barrado, E. M., Fernandez, J., Fernandez, L.(2000). Temporal evolution of groundwater composition in an alluvial aquifer Pisuerga River. Spain by principal component analysis." *Water Research*. Vol. 34. No. 3. pp. 807-816. 2000.

Kaiser, H.F., (1960). The Application of Electronic Computers to Factor Analysis. *Edu. Psychol, Meas* 20: 141-151.

Kim, Y., Lee, K.S., Koh, D.C., Lee, D.H., Lee, S.G., Park. W.B., Koh, G.W., Woo, N.C, (2003). Hydrogeochemical and isotopic evidence of groundwater salinization in a coastal aquifer: a case study in Jeju volcanic island. Korea. *Journal of Hydrology* 270: 282-294.

Lis, G., Wassenaarm, LI., Hendry, MJ. (2007). High-Precision Laser Spectroscopy D/H and 18O/16O Measurements of Microliter Narturel Water Samples. *Anal.Chem*. 80(1): 287-293.

Liu, C. W., Lin, K. H. and Kuo, Y. M. (2003). Application of factor analysis in the assessment of groundwater quality in a blackfoot disease area in Taiwan. *The Science of the Total Environment*. Vol. 313. No. 1- 3. pp. 77-89. 2003.

Mamou, A. (1990). Caractéristiques. Evaluation. Gestion des ressources en eau du Sud Tunisien. PhD thesis. University of Paris South. Paris.

OSS (2003) Système aquifère du Sahara Septentrional Une conscience de bassin. Volume II: Hydrogéologie. *Observatoire du Sahara et du Sahel*. 322.

Pulido- Leboeuf, P. (2004). Seawater intrusion and associated processes in a small coastal complex aquifer (Castell de Ferro. Spain). *Applied Geochemistry*. 19: 1517-1527.

Pulido- Leboeuf, P., Pulido-Bosch, A., Calvache, M.L., Vallejos A., Andreu, J.M. (2003) Strontium. SO_4^{2-}/Cl^- and Mg^{2+}/Ca^{2+} ratios as tracers for the evolution of sea water into coastal aquifers: The example of Castell de Ferro-aquifer. *C.R.Geoscience*. 335: 1039-1048.

Reghunath, R., Sreedhara Murthy, T.R. and Raghavan, B.R.. (2002). The utility of multivariate statistical techniques in hydrogeochemical studies: an example from Karnataka. India. *Water Research*. 36:2437-2442. 2002.

Sânchez-Martos, F.. Jiménez-Espinosa, R.. Pulido-Bosch, A. (2001). Mapping groundwater quality variables using PCA and geostatistics: a case study of Bajo Andarax. southeastern Spain. *Hydrological Sciences-Journal-des Sciences Hydrologiques*. 46(2) April 2001.

Trabelsi, R. (2009). Caractérisation Hydrogéologique et Geochimique du systeme aquifere de la Djeffara (sud-est tunisien): Modélisation et Intrusion marine. PhD Thesis National School of Engineers of Sfax Tunisia 219p.

Trabelsi, R., Kacem, A., Zouari, K., Rozanski, K. (2009). Quantifying regional groundwater flow between Continental Intercalaire and Djeffara aquifers in Southern Tunisia using isotope methods. *Environmental Geology,* 2009, 58 pp.171-183.

WHO (World Health Organization) (1993). Guidelines for drinking water quality. vol.1. *Recommendations* (2nd edn.). Geneva. WHO. pp 130.

Zeng, X., Rasmussen, T. C. (2005). Multivariate statistical characterization of water quality in lake Lanier. Georgia. USA. *J. Environ. Qual*. 34 (6). 2005. 1980-1991.

Zouari, K. (1988). Géochimie et sédimentologie des dépôts continentaux d'origine aquatique du Quaternaire supérieur du Sud Tunisien. PhD thesis, University of Paris-Sud, Orsay.

Zouari, K. Chkir, N. Ouda, B., (2003) Palaeoclimatic variation in Meknassi basin (Central Tunisia) during Holocene period using multidisciplinary approaches. *IAEA,* Vienna, 80-28.

In: Aquifers: Types, Impacts and Conservation ISBN: 978-1-61942-091-5
Editors: Z. Ouakili, et.al, pp. 107-133

Chapter 6

EFFECTS OF HETEROGENEITY ON THE SURFACTANT-ENHANCED REMEDIATION OF AQUIFER CONTAMINATED WITH NONAQUEOUS PHASE LIQUIDS

Kun Sang Lee*
Dept. of Natural Resources and Environmental Engineering
Hanyang University, Seoul 133-791, S. Korea

ABSTRACT

Surfactant-enhanced aquifer remediation (SEAR) is an innovative technology for the removal of nonaqueous phase liquids (NAPLs) in the aquifer. Heterogeneity in aquifer permeability constitutes one of the greatest challenges to remediation process because removal of NAPLs can occur only at the contacted formation by injected surfactant solution. A three-dimensional, multiphase, multicomponent compositional simulator is employed to simulate the solubilization and mobilization of NAPL during the remediation in spatially-correlated heterogeneous fields. Aquifer heterogeneity was accounted for by considering the permeability as a spatial random variable and a geostatistical method was used to generate random distributions of the permeability. Spatial distributions of NAPL saturations in the aquifer

* (Phone) 82-2-2220-2240; (Fax) 82-2-2281-7769; (E-mail) kunslee@hanyang.ac.kr.

and temporal changes of organic recovery, effluent concentrations of organics and surfactant, and pressure drop at the injection well were compared with those in homogeneous aquifer and related to heterogeneity. Variations in permeability fields have a pronounced effect on the organic recovery efficiency due to the long-term persistence of nonaqueous phase liquid in bypassed portion of aquifer and additional dispersion. Higher heterogeneity also results in tailing of effluent organic concentrations and significant loss in injectivity over the remediation life. For a small slug, surfactant-enhanced remediation had a relatively small improvement on the recovery of NAPL, especially in highly heterogeneous aquifers. Migration of the high concentration organic plumes to the other layer by crossflow is also found to have a significant influence on SEAR behavior.

Keywords: Nonaqueous phase liquid, Surfactant-enhanced aquifer remediation, Heterogeneity, Geostatistics

INTRODUCTION

Subsurface contamination by the organic compounds is a complex process and difficult to treat due to many reasons like the tendency of adsorption of contaminants onto the soil matrix, low water solubility, and a limited rate of mass transfer for biodegradation. As many organic compounds have low solubility in water, so they may leach from the soil for a longer period of time and thus ultimately may become a long-term continuous source of the soil and groundwater contamination. Referring to water as the aqueous phase, the liquid organic contaminants remaining as a separate phase in aqueous medium are called nonaqueous phase liquids (NAPLs). Sharp interfaces are maintained between a NAPL and the aqueous phase within each pore.

Conventional remediation method of pump and treat has a limited success in long term remediation of sites contaminated with pure phase NAPLs [1]. Recently, surfactant-enhanced aquifer remediation (SEAR), which is implemented by the injection of dilute aqueous surfactant solutions into contaminated aquifer, has been proposed and refined over the last 20 years as a novel method for the enhancement of pump-and-treat aquifer remediation performance [2, 3]. During SEAR application, the interfacial tension between the injected surfactant solution and the nonaqueous phase contaminants is reduced. Several factors can influence the complex physical and chemical processes affecting the efficiency of aquifer remediation with surfactant.

Among them, geologic heterogeneities are probably the key factors which reduce the effectiveness of chemical enhanced recovery processes because their success depends on the delivery of injected chemical and water into the subsurface to contact the organic liquids. Results of previous studies have indicated NAPL transport are highly sensitive to aquifer heterogeneities and entrapped organic chemicals tend to persist in low permeability regions of heterogeneous formations [4-6].

Further work is needed to form modeling approaches which capture the behavior of the solubilization and mobilization processes and the potential influence of complex, heterogeneous field conditions on SEAR performance. In previous studies, the authors used numerical simulation techniques to predict the distribution of a typical NAPL in a 2D model aquifer under a variety of conditions and for realizations of the underlying permeability field [4, 5]. To date, relatively little attention has been focused on the injectivity of surfactant solution during the process. The study extends those works by describing a systematic study of the influence of structural heterogeneity on the surfactant-enhanced solubilization and mobilization of NAPL in the 3D model setting. A comprehensive numerical model for SEAR was applied to characterize NAPL transport and explore the potential performance in homogeneous and heterogeneous aquifers. Based on the simulated results, a stochastic analysis is used to assess the effect of soil spatial variability and heterogeneity on NAPL recovery and injection pressure as well as spatial and temporal distributions during the remediation process.

MATHEMATICAL FORMULATION

To examine the influence of permeability heterogeneity on SEAR performance, a three-dimensional remediation scenario is considered in which a NAPL is flushed with surfactant solution followed by water in a saturated subsurface aquifer system. Among the most advanced SEAR simulators, a general simulator named UTCHEM has proved to be particularly useful for modeling multicomponent and multiphase transport processes [7]. UTCHEM was originally developed to model surfactant-enhanced oil recovery and later modified for applications involving the use of surfactant for the remediation of aquifers contaminated by NAPLs [3]. UTCHEM has been extensively verified by comparing to analytical and numerical solutions and experimental measurements for its ability to predict the flow of fluids through the aquifer [4, 9-12].

Delshad *et al.* [13] provided detailed information about formulation for modeling surfactant-enhanced remediation, including mass conservation equations, energy conservation equations, pressure equations, equations of nonequilibrium dissolution of NAPL, well models, fluid and soil properties, adsorption of surfactant and organics, cation exchange, phase behavior and phase saturation, and equations of capillary pressure and relative permeability. The formulation for characterizing surfactant-enhanced remediation has been implemented into UTCHEM, a three-dimensional, multicomponent, multiphase, compositional model of chemical flooding processes.

Flow and transport of an organic liquid in a water saturated porous medium can be described by mass balance equations, an overall balance that determines the pressure for up to three phases. The equations incorporate a modified Darcy's law expression for the specific discharge. Under assumptions of local thermodynamic equilibrium, immobile solid phase, slightly compressible aquifer and fluids, Fickian dispersion, and ideal mixing, the basic mass conservation equation for components can be written as follows: [13, 14]

$$\frac{\partial}{\partial t}\left(\phi \widetilde{C}_{\kappa} \rho_{\kappa}\right)+\nabla \cdot\left[\sum_{l=1}^{n_p} \rho_{\kappa}\left(C_{\kappa l} \mathbf{u}_l-\mathbf{D}_{\kappa l}\right)\right]=0 \tag{1}$$

where κ is component index including water (κ=1), organic (κ=2), and surfactant (κ=3); l is phase index including water (l=1), oleic (l=2), and microsemulsion (l=3); ϕ is porosity; $\widetilde{C}_{\kappa}$ is overall concentration of component κ (volume fraction); ρ_{κ} is density of component κ [ML^{-3}]; n_p is the number of phases; $C_{\kappa l}$ is concentration of component κ in phase l (volume fraction); $\mathbf{u}_l$ is Darcy velocity of phase l [LT^{-1}]; S_l is saturation of phase l; R_{κ} is total source/sink term for component κ (volume of component κ per unit volume of porous media per unit time); $\mathbf{D}_{\kappa l}$ is dispersion tensor. The overall concentration ($\widetilde{C}_{\kappa}$) denotes the volume of component κ summed over all phases.

Viscosity

Liquid phase viscosities are modeled in terms of pure component viscosities and the phase concentrations of the organic NAPL, water and surfactant:

$$\mu_l = C_{1l}\mu_w e^{\alpha_1(C_{2l}+C_{3l})} + C_{2l}\mu_o e^{\alpha_2(C_{1l}+C_{3l})} + C_{3l}\alpha_3 e^{\alpha_4(C_{1l}+\alpha_5 C_{2l})} \tag{2}$$

where μ_w and μ_o are pure water and organic NAPL phase viscosities and the α parameters are determined by matching laboratory microemulsion viscosities at several compositions.

Density

Phase specific weights ($\gamma_l = \rho_l g$) are modeled as a function of pressure and composition as follows

$$\gamma_l = C_{1l}\gamma_{1l} + C_{2l}\gamma_{2l} + C_{3l}\gamma_{3l} \tag{3}$$

where $\gamma_{\kappa l} = \gamma_{\kappa R}\left[1 + C_\kappa^o (p_l - p_{Ro})\right]$. $\gamma_{\kappa R}$ and C_κ^o are the component specific weight and compressibility at a reference pressure (p_{Ro}).

Interfacial Tension

For calculating microemulsion/oil (σ_{23}) and microemulsion/water (σ_{13}) interfacial tension (IFT), the IFTs for water and oil (σ_{21}) is assumed to be a known constant. Once the phase compositions have been determined, the interfacial tensions between microemulsion and the excess phases are calculated as functions of solubilization parameters $R_{l3} = \dfrac{C_{l3}}{C_{33}}$.

Trapping Number

The forces controlling the movement of the entrapped blob are the viscous force due to the hydraulic gradient, the trapping force due to capillary pressure and the gravity force, which can act as either a driving or trapping force depending on the direction of the flow. Buoyancy forces can also affect the mobilization of a trapped organic phase. The condition for mobilizing a trapped blob of length L is as follows

$$\Delta L\left|\nabla\Phi_w - g\Delta\rho\right| \geq \Delta p_c \tag{4}$$

The trapping number is defined by the left-hand side of Eq. (4) as

$$N_{Tl} = \frac{\left|-\mathbf{k}\cdot\nabla\Phi_{l'} - \mathbf{k}\cdot\left[g\left(\rho_{l'} - \rho_l\right)\nabla h\right]\right|}{\sigma_{ll'}} \tag{5}$$

Residual saturations are then computed as a function of trapping number as

$$S_{lr} = \min\left(S_l, S_{lr}^{\text{high}} + \frac{S_{lr}^{\text{low}} - S_{lr}^{\text{high}}}{1 + T_l N_{Tl}}\right) \tag{6}$$

where T_l is a positive input parameter based on the experimental observation of the relation between residual saturations and trapping number. S_{lr}^{low} and S_{lr}^{high} are the input residual saturations for phase l at low and high trapping numbers. The endpoints and exponents in relative permeability functions and capillary pressure curves are computed as a linear interpolation between the given input values at low and high trapping numbers ($k_{rl}^{o\text{low}}$, $k_{rl}^{o\text{high}}$, n_l^{low}, n_l^{high})

NUMERICAL MODELING

To explore the factors which could possibly impact the performance of SEAR in heterogeneous aquifer, a hypothetical remediation scenario was devised for model application. Numerical simulation runs were conducted in a three-dimensional domain that includes the aquifer height, i.e., gravity and capillary forces are simultaneously considered. The modeled system used in this study is a square aquifer of 150×150×4 ft^3 and discretized into 15×15×2 grid blocks, which corresponds to $\Delta x = \Delta y = 10\,\text{ft}$ and $\Delta z = 2\,\text{ft}$. Accounting for the spatial variabilities and dimensions in the horizontal and vertical directions of a stratified aquifer, we consider a two-layer system in the vertical direction. The outer boundary is represented as a noflow. Input parameters for the simulations are those that define the physical properties of aquifer, fluid properties, and chemical properties, as given in Table 1.

The hydraulic properties of aquifers have long been recognized as stochastic quantities, or quantities which vary spatially with an element of randomness. To characterize the soil spatial variability and heterogeneity, we treated the permeability of heterogeneous aquifers as a spatial random variable. Unconsolidated aquifer formations have been observed to exhibit strong spatial correlation structure in permeability [15]. It was assumed that permeability values followed a lognormal distribution and its spatial structure was characterized by an exponential variogram.

The assignment of permeabilities from a log-normal distribution is based on the random permeability fields generated by conditional simulation. The parameters are (1) mean of the horizontal permeability ($\bar{k}_h$), (2) variance of logarithm of permeability ($\sigma^2(\ln k)$), and (3) correlation length in x and y directions (λ_x, λ_y). For the same set of values of mean and correlation lengths, different realizations of the random permeability fields were generated by setting different variance ranging from 0.0 to 2.0 and the same seed value for the random number generator in the program. It is also assumed that the random permeability field of the second layer has the same structure with the first layer and was generated by using the same seed number. For each realization of the random variable of aquifer permeability, the generated permeability fields were incorporated into UTCHEM to simulate NAPL transport in heterogeneous aquifers.

Table 1. Properties of aquifer rock and fluids used in the simulations

<table>
<tr><td rowspan="4">rock</td><td colspan="3">rock compressibility (c_r)</td><td colspan="2">0 psi^{-1}</td></tr>
<tr><td colspan="3">porosity (ϕ)</td><td colspan="2">0.2</td></tr>
<tr><td colspan="2" rowspan="2">permeability (k)</td><td>horizontal (k_h)</td><td colspan="2">500 md, 100 md</td></tr>
<tr><td>vertical (k_v)</td><td colspan="2">0.0 k_h - 1.0 k_h</td></tr>
<tr><td rowspan="22">fluid</td><td colspan="3">interfacial tension ($\log_{10}\sigma_{21}$)</td><td colspan="2">1.3 dynes/cm</td></tr>
<tr><td rowspan="12">residual saturation, relative permeability</td><td rowspan="6">low capillary number</td><td rowspan="3">residual saturation (S_r)</td><td>water</td><td>0.37</td></tr>
<tr><td>oil</td><td>0.35</td></tr>
<tr><td>microemulsion</td><td>0.37</td></tr>
<tr><td rowspan="3">relative permeability endpoint (k_r^o)</td><td>water</td><td>0.11</td></tr>
<tr><td>oil</td><td>0.95</td></tr>
<tr><td>microemulsion</td><td>0.11</td></tr>
<tr><td rowspan="6">high capillary number</td><td rowspan="3">residual saturation (S_r)</td><td>water</td><td>0.</td></tr>
<tr><td>oil</td><td>0.</td></tr>
<tr><td>microemulsion</td><td>0.</td></tr>
<tr><td rowspan="3">relative permeability endpoint (k_r^o)</td><td>water</td><td>1.0</td></tr>
<tr><td>oil</td><td>1.0</td></tr>
<tr><td>microemulsion</td><td>1.0</td></tr>
<tr><td colspan="2" rowspan="7">viscosity (μ_l)</td><td>water</td><td colspan="2">0.86 cp</td></tr>
<tr><td>organic</td><td colspan="2">4.00 cp</td></tr>
<tr><td rowspan="5">compositional phase viscosity parameters for microemulsion viscosity</td><td>α_1</td><td>2.5</td></tr>
<tr><td>α_2</td><td>2.5</td></tr>
<tr><td>α_3</td><td>10.0</td></tr>
<tr><td>α_4</td><td>1.0</td></tr>
<tr><td>α_5</td><td>1.0</td></tr>
<tr><td colspan="2" rowspan="2">density (ρ_l)</td><td>water</td><td colspan="2">0.433 psi/ft</td></tr>
<tr><td>organic</td><td colspan="2">0.368 psi/ft</td></tr>
</table>

The mean permeability of each layer was set equal to the permeability of the homogeneous aquifer, that is, 500 md and 100 md. Four values of variance of log of permeability ($\ln k$), that is, 0.0 (homogeneous), 0.5, 1.0, and 2.0

were selected to generate random fields of the soil permeability. The choice of the standard deviations of $\ln k$ was made by considering the common heterogeneity of soils [16, 17]. The correlation lengths in the x and y directions were chosen as 30 m. A sample realization of an associated lognormally-distributed permeability field for $\sigma^2(\ln k) = 0.5$ is presented as Figure 1. The results from flooding the homogeneous reservoir were chosen as a comparison standard to identify the zero level of formation variability. Further simulations were performed for a more heterogeneous cases to examine how model sensitivity to input parameters changes in systems with higher variance of the permeability field.

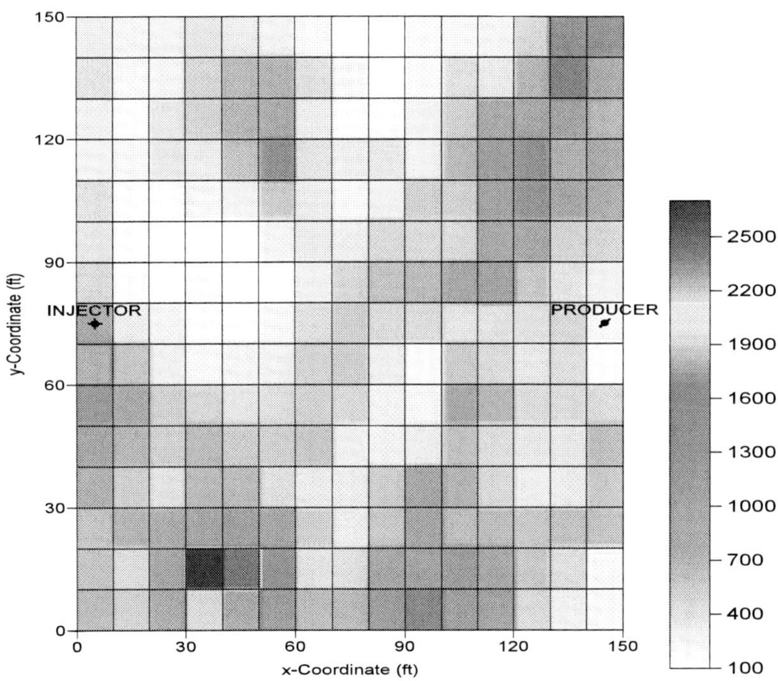

Figure 1. Permeability distribution of the first layer in randomly heterogeneous aquifer with $\sigma^2(\ln k) = 0.5$.

To explore the behavior of a typical remediation process in a heterogeneous aquifer, a base simulation scenario was defined describing a SEAR in a field-scale domain. An injection well is located at the center of the left-hand side of the x–y plane, with x and y coordinates (ft) of (7.5, 75.0). A production well is at the right-hand side with coordinates (ft) of (142.5, 75.0). All the wells are vertical and completed in the z direction. A constant injection rate of 50 ft^3/day was used during the remediation processes. The initial relative saturations of water and NAPL were assumed to be 0.35 and 0.65, respectively, uniformly distributed in the contaminated region at the center of the first layer. The uncontaminated region is saturated with water. The remediation process was as follows: (1) inject of surfactant solution for a certain time at a constant rate, and (2) inject water at the same rate. The simulations were conducted over a 700-day period.

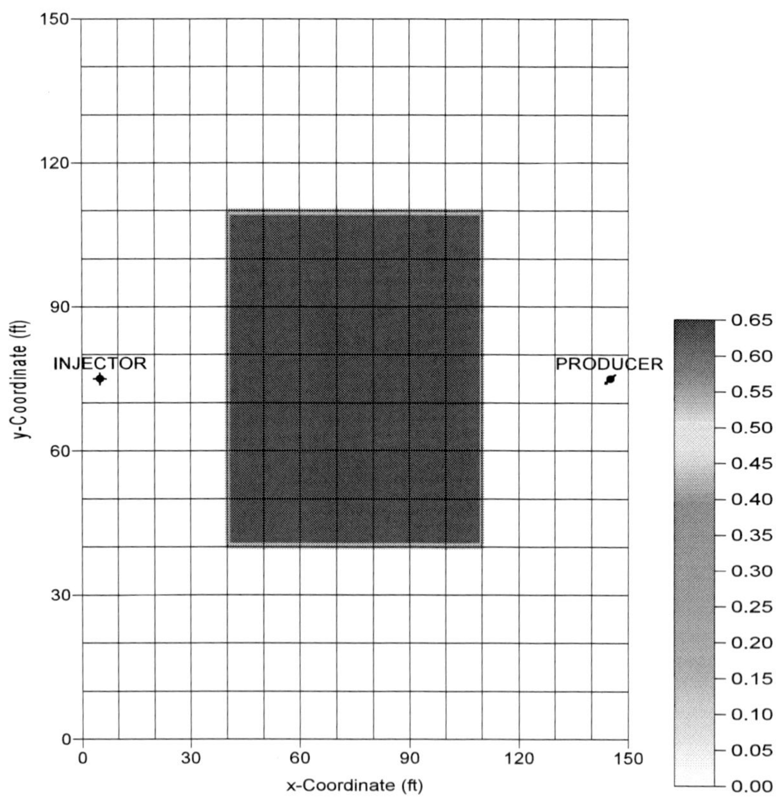

Figure 2. Initial distribution of NAPL in the domain.

The SEAR injection sequence for the cleanup of the NAPL spill consists of surfactant slug introduced by 3% solution and chase water. The surfactant solution passes through the zone of contamination at the center of the first layer. Figure 2 depicts initial distribution of organic liquid which are used to illustrate the performance of the remediation processes. In the following analyses, we used the simulated results based on an initially uniform NAPL distribution in homogeneous and heterogeneous soils.

RESULTS AND DISCUSSION

Heterogeneity

For each of the four values of variance of $\ln k$, randomly distributed permeability fields were generated. Each of the random permeability fields was incorporated with UTCHEM for simulations of surfactant-enhanced remediation in heterogeneous aquifers. A 3% surfactant solution was injected at a constant injection rate of 50 ft^3/day for 150 days, followed by chase water for the remaining remediation period. The process was conducted on the heterogeneous reservoirs in an identical fashion so any difference in organic recovery can be attributed to the differences in permeability heterogeneity. The simulated results were then used to analyze effects of aquifer spatial variability and heterogeneity on NAPL transport in the aquifer and recovery from a production well.

A graphical depiction is presented in Figure 3 for an organic distribution after 700 days of the surfactant-enhanced removal of the organic liquid in a formation having $\sigma^2(\ln k)$ of 0.0, 0.5, 1.0, and 2.0. It can be seen that a significant portion of the organic liquid initially located at the center of the aquifer has been removed, leaving more isolated regions of entrapped NAPL in areas of lower permeability for the case of heterogeneous aquifer. The figure reveals that the presence of heterogeneity increases the variability in NAPL saturation, resulting in an increase in maximum organic saturation from 0.31 for the homogeneous case to 0.36 for $\sigma^2(\ln k) = 2.0$, which indicates a larger variability in S_o after remediation. Simulator output for NAPL saturation in the higher variance case shows a visible increase in lateral spreading relative to that of homogeneous case.

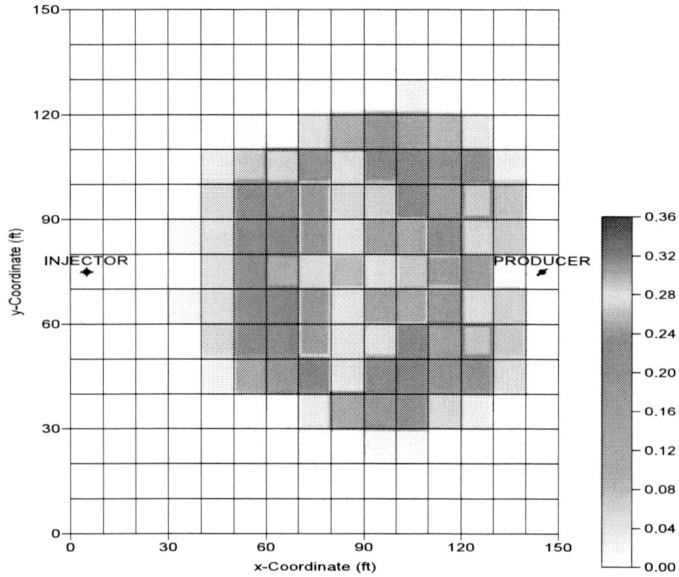

(a) $\sigma^2(\ln k) = 0$

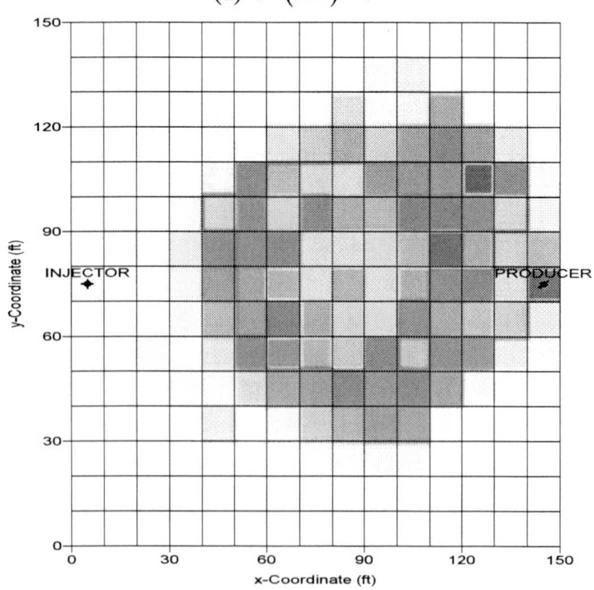

(b) $\sigma^2(\ln k) = 0.5$

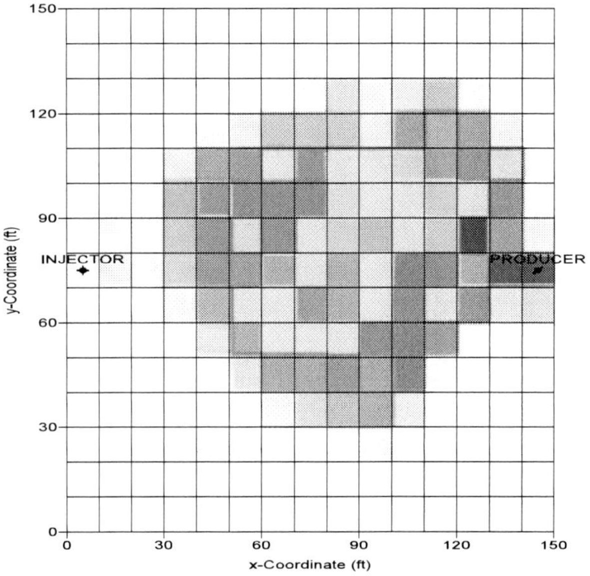

(c) $\sigma^2(\ln k) = 1.0$

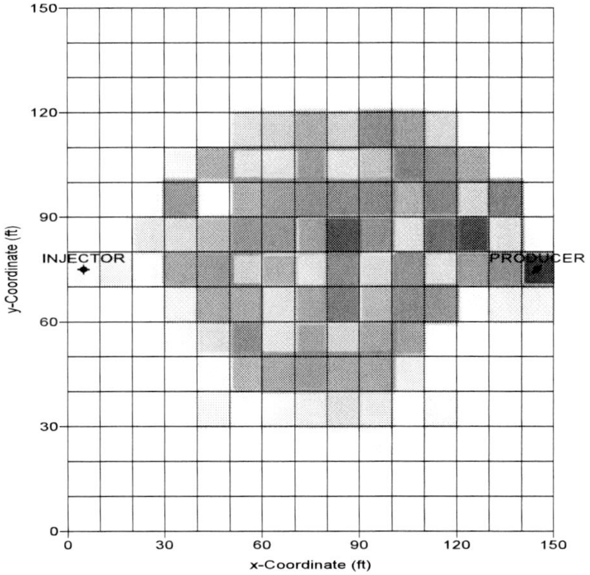

(d) $\sigma^2(\ln k) = 2.0$

Figure 3. NAPL saturation in the first layer after flushing for 700 days.

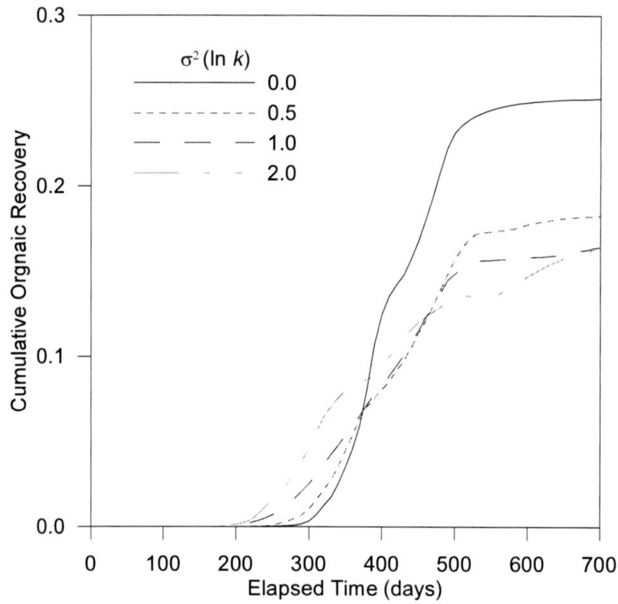

(a) cumulative organic recovery

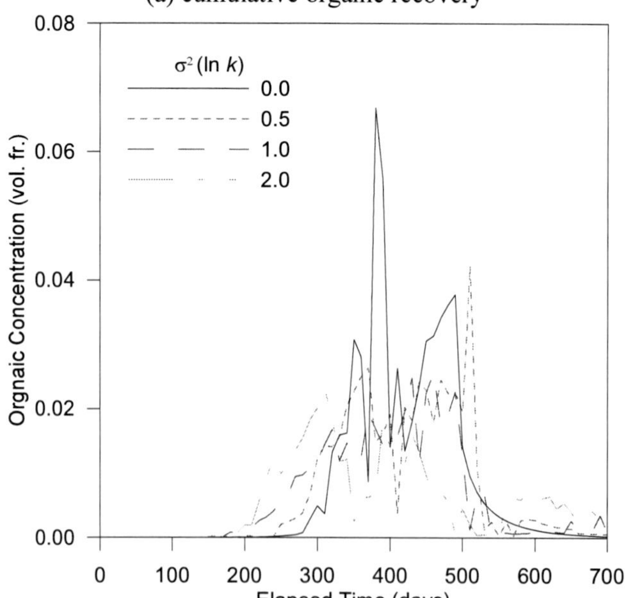

(b) concentration of recovered organic chemicals

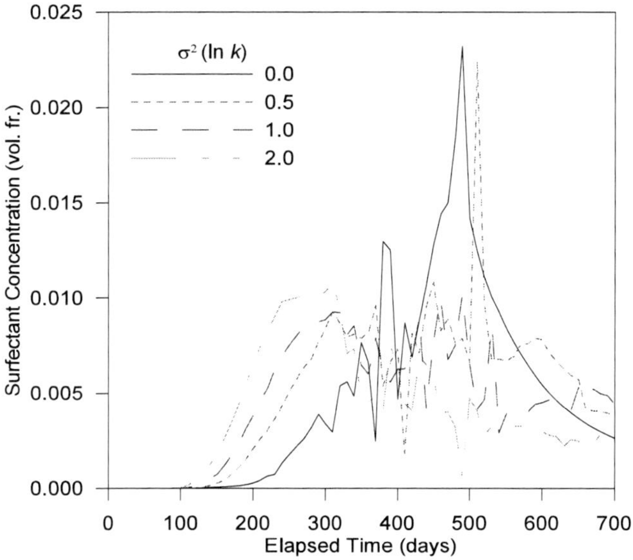

(c) concentration of recovered surfactant

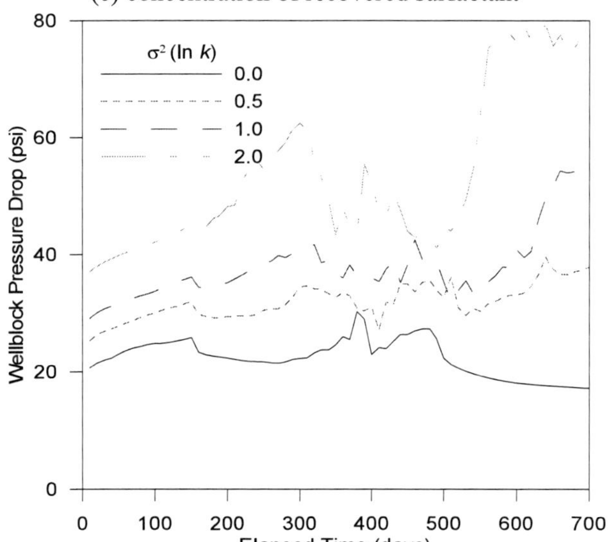

(d) wellblock pressure drop

Figure 4. History of injection and production wells obtained from simulations with different level of heterogeneity.

The results of simulations at the four levels of variability in $\ln k$ are presented in Figure 4. The simulator predicts NAPL recovery and wellblock pressure drop for aquifers having the identical mean conductivity and spatial correlation structure, but various levels of heterogeneity in permeability field. As the aquifer heterogeneity (characterized by the values of $\sigma^2(\ln k)$) increases, the organic recovery is greater due to channeling along the high permeability region at early time. However, cumulative NAPL recovery decreases from 0.25 to 0.16 at 700 days. The reduction of NAPL recovery is attributed to the reduction of sweep efficiency in heterogeneous aquifers. As the heterogeneity increases, variation of the permeability also increases with high and low values within a small region, which results in large variation of flow velocity in the region. The contrasts in the permeability cause channeling and flow barriers, which results in early breakthrough of injecting fluid and a lower sweep efficiency. The additional dispersion caused by the randomly heterogeneous reservoir resulted in less NAPL recovery than that for the homogeneous reservoir. The increase of $\sigma^2(\ln k)$ up to 2 in surfactant process caused the production of NAPL to be 35% less than that of homogeneous case.

The calculated concentrations of dissolved NAPL and recovered surfactant in all phases from the production well are plotted as a function of time in Figures 4(b) and (c). The surfactant front is reaching the region of entrapped NAPL, resulting in increased solubilized concentration in the aqueous phase, then recovered at the producing well. Therefore, the concentrations of surfactant and organic chemicals have a similar trend with some delay. In the case of homogeneous aquifer, the calculated NAPL and surfactant concentrations show a relatively clear high peak and dramatically decreases as NAPL mass is depleted. They also indicate the early breakthrough of surfactant front and NAPL along high permeability path and an associated tailing of effluent organic concentrations at the production well in heterogeneous aquifers. The breakthrough of NAPL in the aquifer with $\sigma^2(\ln k) = 2.0$ occurs around 140 days, whereas is about 140 days earlier than that from the homogeneous case. It is not possible for surfactant to reach all of the entrapment zones in heterogeneous aquifers, and some locations of NAPL entrapment may undergo slow remediation with a given injection sequence. As surfactant delivery and NAPL recovery is controlled by the heterogeneity of the aquifer, the concentration curves have much lower peaks.

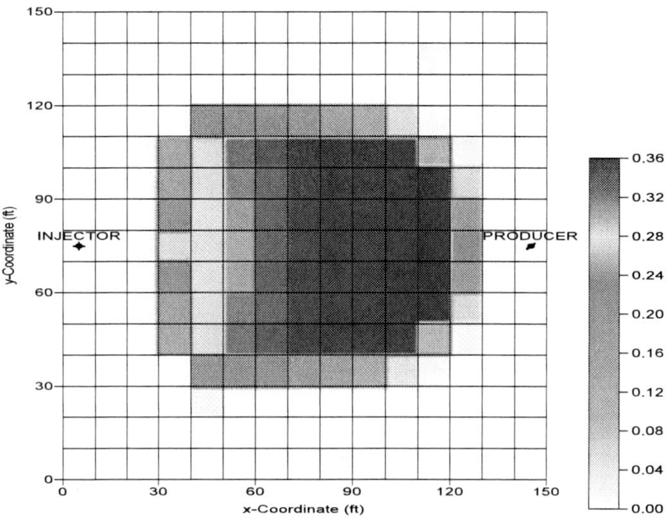

(a) $t_{inj} = 10$ days

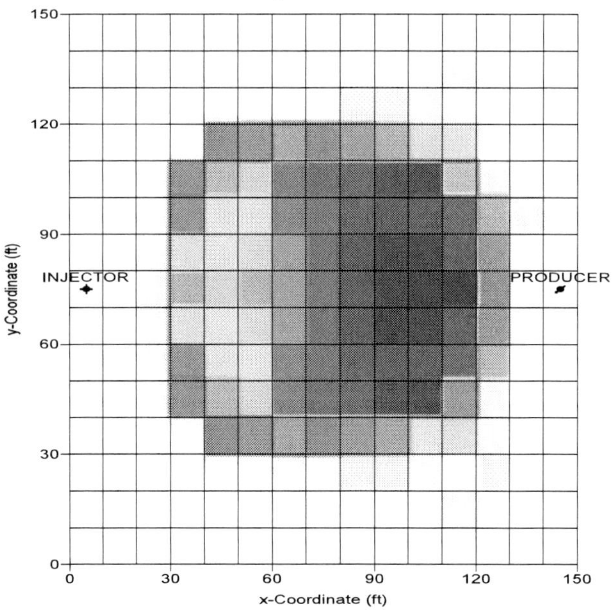

(b) $t_{inj} = 25$ days

Figure 5. (Continued)

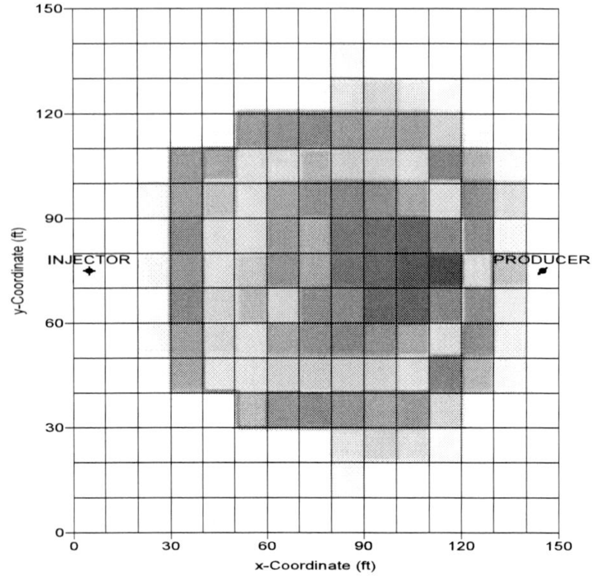

(c) $t_{inj} = 50$ days

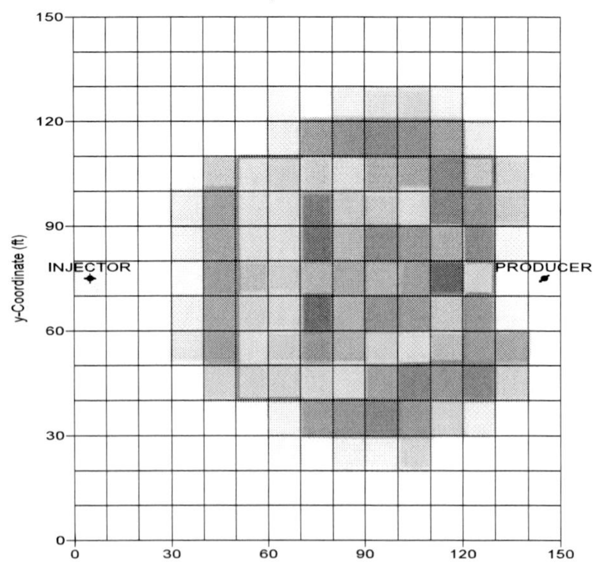

(d) $t_{inj} = 100$ days

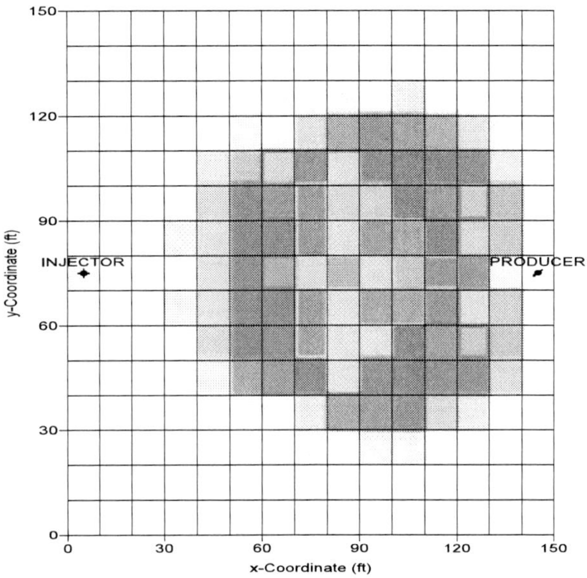

(e) $t_{\text{inj}} = 150$ days

Figure 5. NAPL saturation in the first layer of homogeneous aquifer after flushing for 700 days with different slug injection time.

Increased variability in the permeability field also results in the loss of injectivity. The comparative results on the pressure drop at the injection well among the aquifers are considerable. Examination of Figure 4(d) reveals that the presence of increasing variability results in a clear increase in the wellblock pressure drop required to maintain the constant injection rate. The highly heterogeneous case showed about 2.45 times greater pressure drop compared to the homogeneous case. A large pressure drop indicates the operational ineffectiveness of using a surfactant flooding for the highly heterogeneous aquifer.

Slug Size

In this section, we study the effect of surfactant slug size on the NAPL recovery efficiency from an aquifer, according to five different injection policies. This study was performed by a series of simulations where all parameters were kept the same as previous runs except the surfactant injection time. The injection time of surfactant solutions was varied from 10 to 150 days, while the injection rate was held constant.

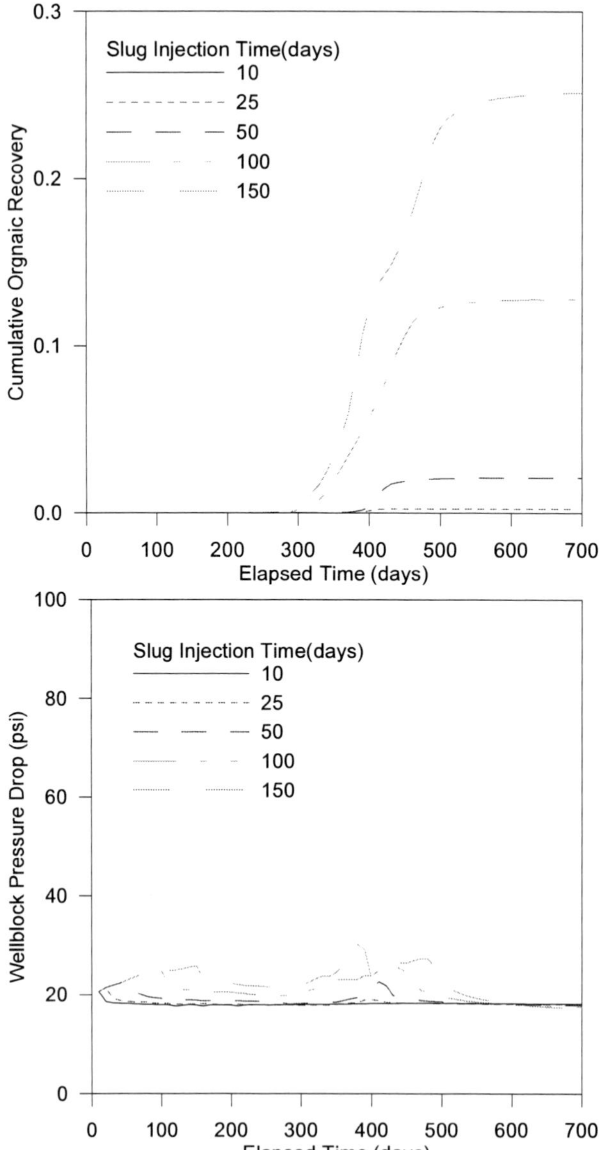

(a) $\sigma^2(\ln k) = 0.0$

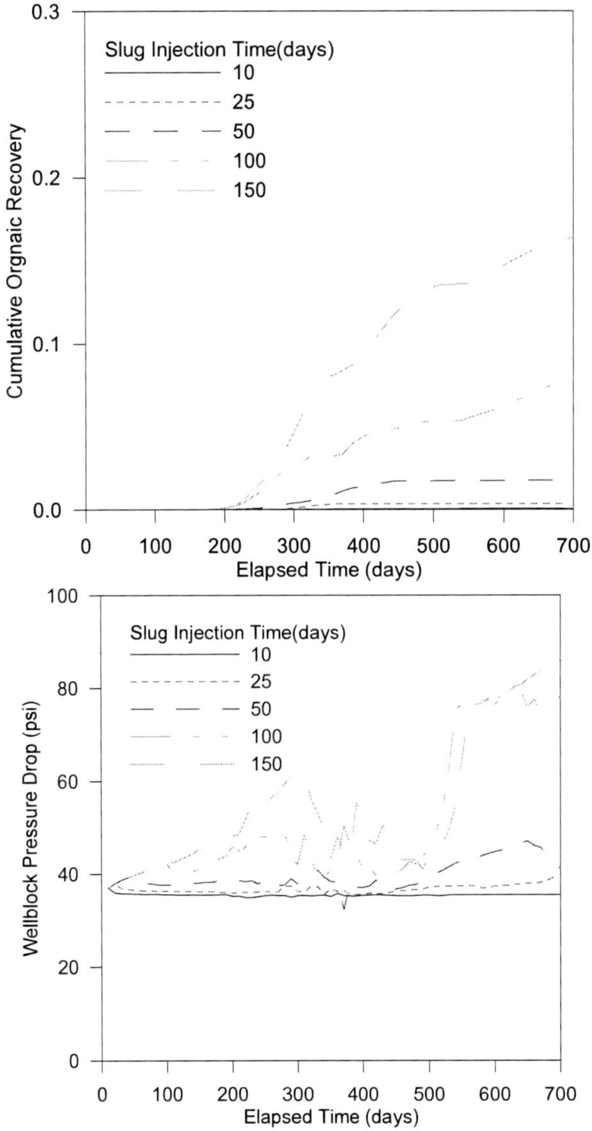

(b) $\sigma^2(\ln k) = 2.0$

Figure 6. History of injection and production wells obtained from simulations with different slug injection time.

Results of the simulations performed are presented in Figures 5 and 6, for both the cases of homogeneous aquifer and highly heterogeneous aquifer of largest variance of $\ln k$. A visual inspection of the plots reveals a clear trend that a surfactant process with larger slug size can significantly enhance NAPL recovery. In some extreme cases with a small slug, the NAPL will not be recovered. This behavior can be explained by the influence of the dispersion of surfactant slug during access to the entrapment zones. Although an increased recovery is obtained with a larger slug, the increase is not proportional to the slug size. The incremental recovery is insignificant at a relatively small slug size, but increases greatly with the increased slug size.

To further explore the relation between slug size and permeability variance, it is useful to compare the results of homogeneous case with predictions for the higher variance formation. Figure 6(b) presents a plot of the recovered organic in the most variable formation. It can be seen that more variability tends to obscure the effect of the slug size. The results suggest that there is a strong effect of heterogeneity on required cleanup volume.

A significantly-increased pressure drop is also obtained by the injection of a larger slug of surfactant solution. Comparison of Figures 6 (a) and (b) shows an increased pressure drop of model predictions at the higher level of variance in $\ln k$. In a homogeneous aquifer, the average pressure drop increases from 18.1 psi to 22.4 psi with increasing slug. The highly heterogeneous case showed average pressure drop of 35.5 psi to 54.0 psi, which are about 1.96 to 2.41 times larger than those of the homogeneous case. In a highly heterogeneous aquifer, the increase in pressure drop due to larger slug size is much clearer.

Crossflow

A number of simulations have been directed toward the evaluation of crossflow in the layered aquifer for the dissolution and mobilization of NAPLs. An anisotropy ratio of permeabilities between horizontal and vertical directions was used to simulate various level of crossflow in the stratified formation. Five cases of surfactant flooding were run in which the ratio of vertical to horizontal permeability (k_v/k_h) was varied from 0.0 to 1.0. Calculations were made for 150 days of surfactant injection. Figures 7(a) and (b) show typical results of surfactant flood in stratified reservoirs on oil recovery and concentration of recovered NAPL.

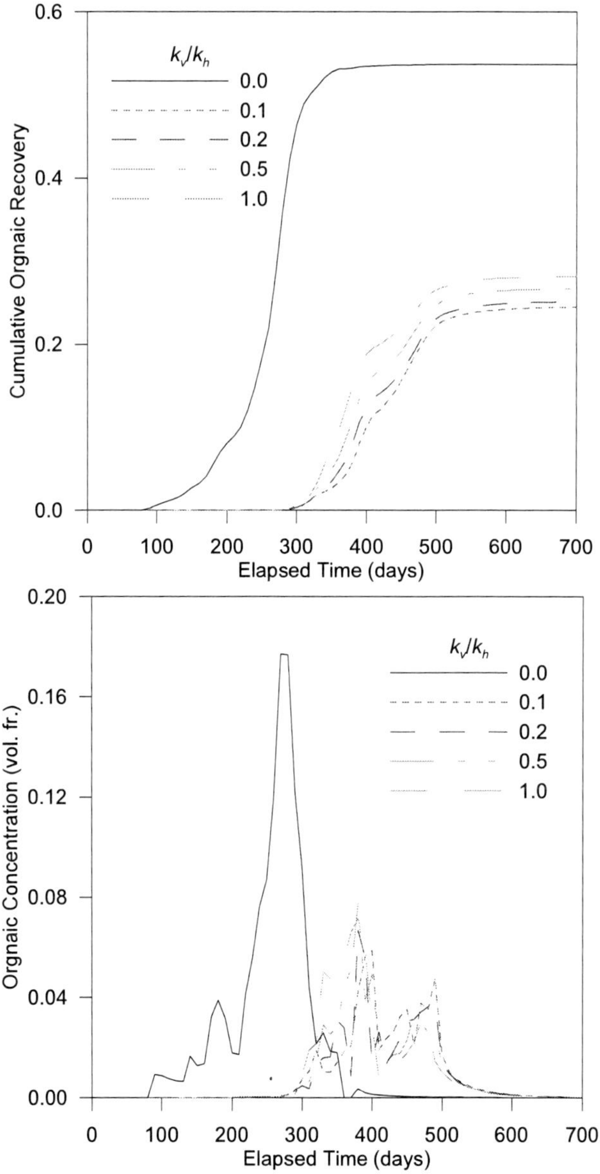

(a) $\sigma^2(\ln k) = 0.0$

Figure 7. (Continued)

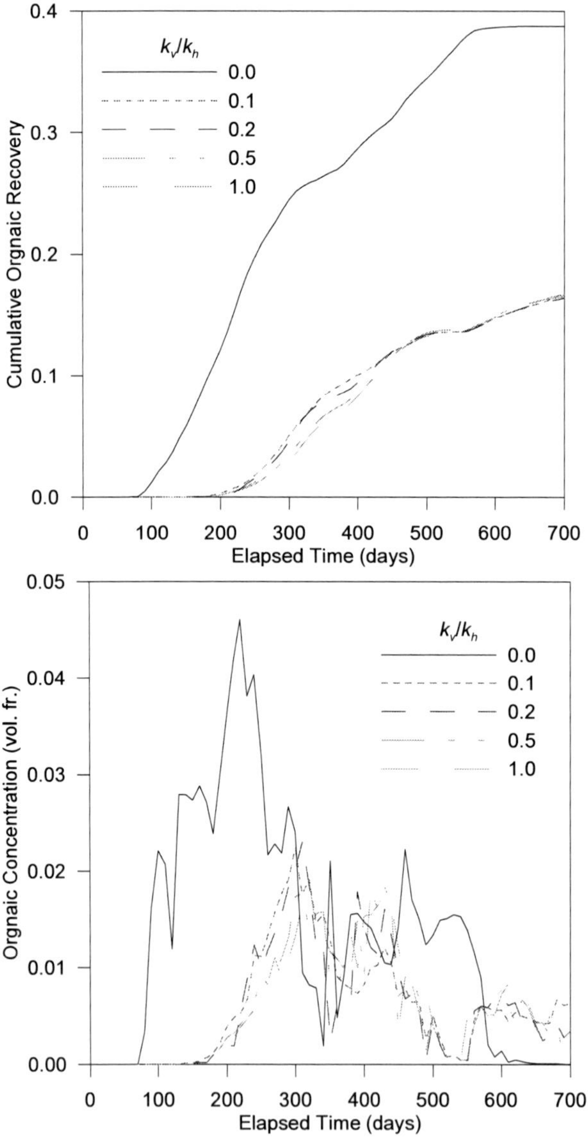

(b) $\sigma^2(\ln k) = 2.0$

Figure 7. History of production well obtained from simulations with different level of crossflow.

It is shown that when polymer solution is injected into a stratified reservoir with layers of differing mean permeability, the NAPL recovery is dominated by vertical crossflow. In the case of no crossflow ($k_v/k_h = 0$), the dramatic increase in the effectiveness of surfactant flood is obtained. Vertical stratification without crossflow reduces the dispersion of NAPL to the other layer and increases displacement efficiency of the flood.

NAPL recovery increases as the vertical permeability increases for the SEAR process in the homogeneous layered aquifer. Sufficient vertical permeability allows polymer to induce crossflow in the aquifer. As the vertical permeability increases, the solubilized organic plume migrates to other layers more easily, where it is swept by the displacing water to the producing well. For heterogeneous aquifers, an increase in the degree of crossflow leads to a slightly decreased NAPL recovery. The presence of same level of variance in $\ln k$ causes the dispersion of NAPL at the other layer after crossflow.

CONCLUSIONS

To assess the potential influence of field-scale formation heterogeneity on the performance of surfactant-enhanced remediation, a 3D modeling study was performed to investigate SEAR behavior for a model system. Random, spatially-correlated fields were generated using the geostatistical method to simulate the permeability structure and the simulator was used to predict spill conformation under a variety of conditions. The simulator was used to model SEAR performance for the range of spill configurations predicted by the multiphase flow simulator, and the sensitivity of model predictions to the model parameters was explored.

The numerical work presented here provides an assessment of the influence of field-scale heterogeneity on the SEAR process in a model saturated aquifer system. The results illustrated the long-term persistence of NAPL, the associated tailing of effluent organic concentrations, and increase in pressure drop at injection well in highly-heterogeneous fields. The effectiveness of SEAR is sensitive to injection time due to dispersion of surfactant slug with the presence of a higher degree of heterogeneity. The tendency of the high concentration organic plumes to migrate to the other layer by gravitational forces and crossflow is also demonstrated and is shown to enhance remediation of aqueous phase organic. Consequently, characterization of permeability in terms of both areal and vertical variability and correlation structure will be crucial to a more complete understanding of the transport and remediation of NAPL in field scenarios.

REFERENCES

[1] Jayanti, S; Pope, GA; Weerasooriya, V; Zhong, L; Dwarakanath, V; Malik, T. SPE 66753, *SPE/EPA/DOE Exploration and Production Environmental Conference*, 2001.
[2] Lowe, DF; Oubre, CL; Ward, CH. *Surfactants and Cosolvents for NAPL Removal*; Lewis Publishers: Boca Raton, FL 1999.
[3] Fiorenza, S; Miller, CA; Oubre, CL; Ward, CH. *NAPL Removal: Surfactants, Foams, and Microemulsions*; Lewis Publishers, Boca Raton, FL 2000.
[4] Brown, CL; Pope, GA; Abriola, LM; Sepehrnoori, K. *Water Resour. Res.* 1994, 30, 2959-2977.
[5] Dekker, TJ; Abriola, LM. *J. Contaminant Hydrology* 2000, 42, 219-251.
[6] Zhang, R; Wood, AL; Enfield, CG; Jeong, S-W. *J. Environ. Qual.* 2003, 32, 957-965.
[7] Center for Petroleum and Geosystems Engineering. *UTCHEM-9.0 A Three-Dimensional Chemical Flood Simulator*; U. of Texas at Austin, Austin, TX, 2000.
[8] Ooi, KC; Dwarakanath, V; Holzmer, FJ; Jackson, RE; Pope, GA; Weerasooriya, V. SPE 52732, *SPE/EPA Exploration and Production Environmental Conference*, 1999.
[9] Ouyang, Y; Cho, JS; Mansell, RS. *Water Res.* 2002, 36, 33-40.
[10] Zhang, J; Delshad, M; Sepehrnoori, K. SPE 94222, *SPE/EPA/DOE Exploration and Production Environmental Conference*, 2005.
[11] Schaerlaekens, J; Mertens, J; Linden, JV; Vermeiren, G; Carmeliet, J; Feyen, J. *J. Contaminant Hydrology* 2006, 86, 176-194.
[12] Qin, XS; Huang, GH; Chakma, A; Che, B; Zeng, GM. *Sci. of the Total Environ.* 2007, 381, 17-37.
[13] Delshad, M; Pope, GA; Sepehrnoori, K. J. *Contaminant Hydrology* 1996, 23, 303-327.
[14] Chen, G; Li, Y; Wang, J; Ma, M; Lu, K; Jin, G; Sun, H. SPE 114346, *SPE/DOE Improved Oil Recovery Symposium*, 2008.
[15] Jussel, P; Stauffer, F; Dracos, T. *Water Resour. Res.* 1994, 30, 1803-1817.
[16] Rehfeld, KR; Gelhar, LW. *Water Resour. Res.* 1992, 28, 2085-2099.

[17] Efendiev, Y; Durlofsky, LJ; Lee SH. *Water Resour. Res.* 2000, 36, 2031-2041.

In: Aquifers: Types, Impacts and Conservation ISBN: 978-1-61942-091-5
Editors: Z. Ouakili, et.al, pp. 135-147

Chapter 7

POLLUTION RISK OF GROUNDWATER, IN A SEMI ARID REGION BY WASTEWATER REJECTIONS: CASE OF TEBESSA AQUIFER SYSTEM

Abdelkader Rouabhia[1*], Moufida Bouteraa[1], Fethi. Baali[1], Chemseddine Fehdi[1], Gérard Vergoten[2]
[1]Department of Earth Sciences, University of Tébessa. Algeria.
[2]Université de Lille. France

ABSTRACT

Groundwater is exposed, more and more intensively, to deliberate discharges of polluting effluents, sewage or storm water runoff in urban areas. Near cities, the sources of groundwater contamination are numerous and are related to many urban activities. The urban water is a source of contamination of groundwater by their concentration in organic and inorganic constituents. Other sources can be added such as air pollution, rain, washing pavements, etc... However, their origins can be complex and include stormwater, wastewater leakage networks. Leachates of municipal waste, septic tank are considered as a source of

---

* Corresponding author: Rouabhia Abdelkader
BP 34-A 12004. Nahda-Poste. Tébessa. Algeria
E mail: rouabhia22@yahoo.fr Telephone: 00 213 772470096

contamination and carry pollutants that have major environmental impacts.

On Oued El Kebir basin in Tébessa (fig1) increased industrial and agricultural activities, and the alternative by-products of production or post-consumer waste, making them vulnerable groundwater resources. The plan amendments leading to variations in groundwater level by excessive pumping in urban areas can also cause contamination from interconnected flows. Impact may be a hydrological since a major exploitation of the resource has an action on the hydrodynamic behaviour of the aquifer.

In this perspective, we try to examine the state of the groundwater resources of Tebessa plain, the last cut by the Oued El Kebir, which drains the wastewater from the entire urban area of the city and which location can contaminate groundwater.

Keywords: Groundwater; Hydrochemistry; Nitrate pollution; Wadi El Kebir, Algeria.

INTRODUCTION

Location

The area of study is near the city of Tebessa in eastern Algeria (Figures 1a, 1b). This area lies in the semi-arid region of Algeria and is susceptible to the various threats common in both growing urban areas and developing agricultural areas. The city of Tebessa and the surrounding villages (Bekkaria, Hammamet) have seen a great deal of growth in the past decade, with the establishment of new industries and farms.

The purpose of this study is two-fold. The first is to study the hydrogeological regime in this area with the help of chemical and isotopic data. Such studies can help constrain hydrogeological regimes, which are poorly covered by water-level data. The second purpose is to evaluate the present status of water quality in the area of study. This will help current water resource planning in the area and will provide a baseline for future studies of water quality and trends. The Tebessa area straddles two aquifers: the limestone aquifer and the alluvial aquifer. In general, the shallow groundwater in Merdja area is found in the alluvial fan deposits from plio-quaternary age (Figure 2). This aquifer overlies geologic formations consisting of cenomanian marly layers. The recharge of this alluvial aquifer is combined and is done from: (1) highlands of Bouroumane on the East and Doukkane on the West (Precipitation infiltrating through alluvial fans where the mountain range meets the plain); (2) from losing streams; and (3) through cross-formational flow.

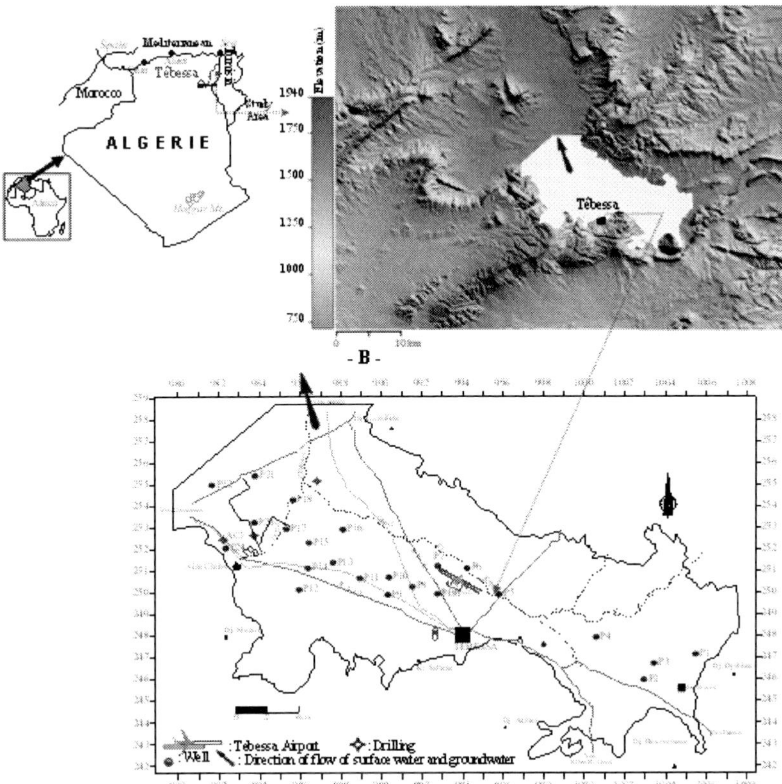

Figure 1. Location of study area in Algeria: A) Topographic Map, B) the Boundaries of the study area.

Climate and Precipitation

Annual precipitation in the studied area ranges from 200 to 350 mm (Figure 3), thus this considered to be a semi-desert area. Summer temperatures can reach 45°C. This situation of dryness accentuates the drawdown of water resource especially during the last decade because the renewal of this resource is very weak. Dryness generates sometimes very unfavorable effects on groundwaters. (Aquifer refill decreases considerably whereas the exploitation increases).

The dry climate, atmospheric dust and low precipitation can also affect the water quality generally causing increased salt content.

Geological and Hydrological Setting

The geology of studied area was investigated by several authors J.L Blés and J.J Fleury (1970), Vila (1977 and 1980). The micropaleontological and biostartigraphical analysis, have showed that from the stratigraphical point of view, the studied area covers the plio-quaternary tectonic depression of Tebessa, This depression separates the highlands of Dyr situated in the North from the one of Doukkane and Mestrie highlands in the South.

The studied area is constituted in the major part by cretaceous formations (Figure2), forming a series of anticlines and synclines. The stratigraphic sequence is presented in the form of alternation between carbonated formations of limestones, marly-limestones and argillaceous marls.

The plio-quaternary and quaternary terrains occupy the central part; they are consisted by actual and recent alluvial deposits, conglomerates, gravels, sandstones…etc.

The summary analysis of the stratigraphic column of the studied area shows the presence of three (03) aquiferous formations among them the formation of plio- quaternary one. This aquifer of great extension occupies the major part of the tectonic depression, limited at the West and at the East by two great faults of NW-SE orientation. It is consisted very varied deposits such as, alluvial fans, silts, calcareous crust, conglomerates and gravels. This aquifer plays a great role in the drinking water supply of the inhabitants of this area.

Hydrochemical Data

Sampling and Analysis

The physico-chemical and chemical data of all the investigated groundwater are summarised in Table 1. According to the analytical data, Parameters values range widely except for the pH and dissolved Oxygen (DO) and nitrate concentrations. DO and pH values increase slightly eastwards up to a maximum value of 8.6 mg/l and 7.92 respectively, whereas nitrate concentration increases westwards, with a maximum of 281 mg/l in December 2008. Significant changes in water quality occur along the river course due to the influence of its tributaries, wastewater disposal, agricultural activities, and large seasonal variations in the discharge rate of the Merdja Plaine. Variations in selected water-quality parameters (EC, NO_3^-, Cl^-, SO_4^{2-},and Na^{2+}) along the Merdja plain are illustrated in Figure 4. In general, water quality becomes considerably worse downstream of the Merdja plain due to a significant increase in SO_4^{2-} (>800 mg/l) and NO_3^- (> 50 mg/l) concentrations. This is attributed to the use of wastewater of Wadi el Kebir in irrigation.

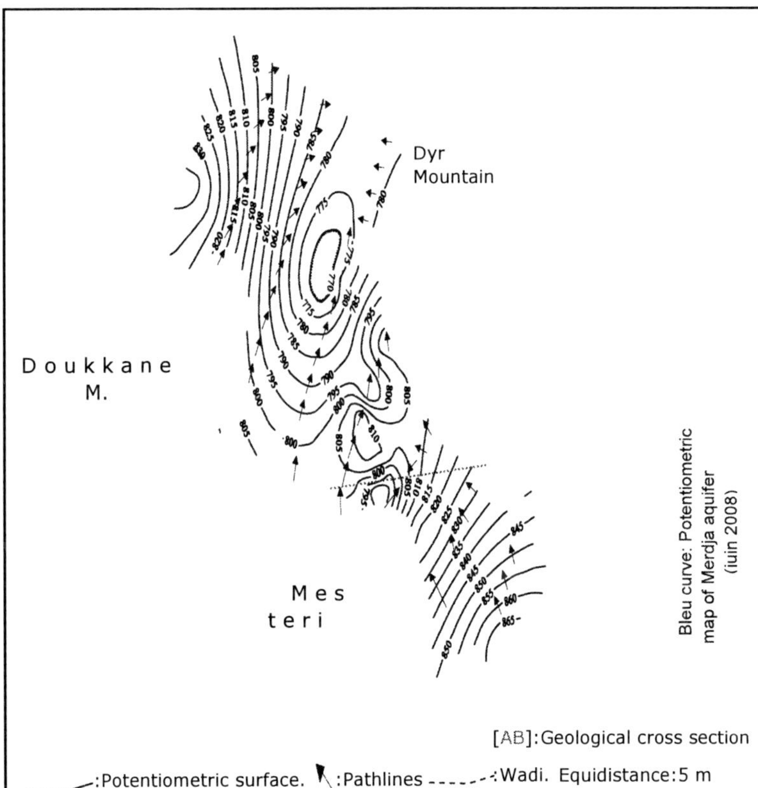

Figure 2. Geological and hydrogeological boundary conditions map of the study area. In Rouabhia 2010.

Table 1. Chemical composition of groundwater from the Merdja aquifer. 2010

well	TDS	Ca^{2+}	Mg^{2+}	Na^{+}	K^{+}	Cl^{-}	SO_4^{2-}	PO_4^{2-}	NO_3^{-}	NO_2^{-}	EC (µS/cm)
						(Mg.L⁻¹)					
1	589	113	229	91,16	5,6	129	159	0.07	28,5	<0.1	682
2	824	0,14	0,46	37,59	138	104	91,6	0.11	11,6	0.12	784
3	881	105	231,6	96,8	4	92	249	1.55	38,4	2.22	1272
4	377	187	228,8	54,51	3,2	97	138	0.07	11,5	0.51	398
5	1071	157	225,7	34,77	6	226	111	0.13	7,3	3.21	2320
6	920	199	233	139,1	3,4	360	255	0.11	13,4	3.66	2682
7	1039	--	175	116,5	4,4	397	350	0.12	18,3	0.40	1324
8	2176	146	233,3	37,59	2,4	221	92	0.23	12,8	1.90	3250
9	922	132	116,7	--	2,4	148	94	0.05	8,6	0.11	530
10	1400	81	234,5	34,77	3,6	223	192	0.27	1,2	0.26	2553
11	676	80	140,1	113,7	3,9	176	350	0.15	31,1	0.30	2859
12	432	188	231	139,1	5,9	97	243	0.07	49,3	0.20	539
13	767	111	232	48,87	4,9	450	112	0.09	8,5	0.30	539
14	627	100	227,4	48,87	2,5	153	67	1.56	9,6	0.40	1606
15	460	159	134,8	65,79	3,8	129	134	0.18	13,4	2.50	1766
16	737	188	232	46,05	4	129	65	0.11	9,9	2.41	2784
17	988	176	130,9	99,62	4,6	177	149	0.19	36	1.14	1428
18	563	300	229	60,15	2,5	115	83	0.1	7,8	0.20	833
19	1079	187	238,5	57,33	2,7	223	76	0.09	5,1	0.60	590
20	571	204	223,5	71,43	4,4	329	49	0.05	8,8	0.60	1031

3. RESULTS

3.1. Hydrochemistry

Results of analysed parameters in groundwater samples are given in Table 1. The quality parameters have wider ranges due to spatial changes in the lithology of the alluvial sediments that constitute the aquifer and seasonal changes in fertilizer application and irrigation practices.

The spatial distribution of the values of EC, NO_3^- and SO_4^{2-} in groundwater in December 2010 shown in Figures 4.1, 4.2 and 4.3 These maps provide a basis for making area wide generalizations concerning the distribution of water quality parameters.

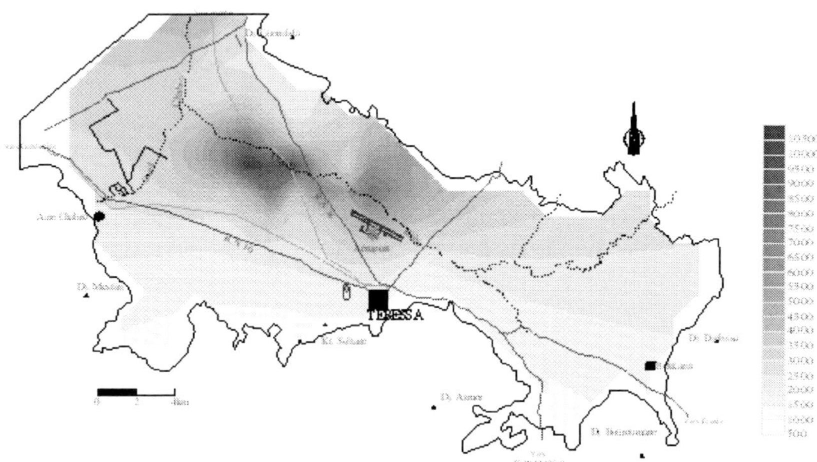

Figure 3.1. Electrical conductivity (EC) distribution in groundwater of the study area.

EC and NO_3^- spatial distribution patterns are very similar, and show that the dissolution of evaporites significantly affect groundwater chemistry. Mention should be made of the high concentrations of SO_4^{2-} (>800 mg/l) and NO_3^- (> 50 mg/l) in wells located between Tébessa and Chabro wadi and downstream of the Ain Chabro.

Other sources of sulfate in groundwater may be agricultural practices, although fertilizers containing sulfate are not commonly used in the study area.

3.2. Water Quality

The purpose of this section is to characterize groundwater for both domestic as well as for irrigation purposes. Water was classified based on total hardness according to the classification of Sawyer and McCarty (1967).

Hardness is defined by the presence of divalent cations of which calcium and magnesium are the most abundant in groundwater.

Total hardness (TH) is calculated through the following equation (Todd 1980)

$$TH = 2.5\ Ca(mg\ l^{-1}) + 4.1\ Mg(mg\ l^{-1})$$

All sampled wells, located in the studied area are characterized as hard except the well M2 located in the eastern part, classified as a very hard water type.

Nitrate is a major contaminant of drinking water. It is currently frequently found in aquifers. In arid and semi-arid regions, sources of nitrate in groundwater have either been linked to direct anthropic pollution in urban areas or to leaching of fertilizers in agricultural areas (Girard and Hillaire 1997). Nitrate (NO_3) concentrations of the phreatic waters (Table 1) were found far above the World Health Organization (WHO 1998) recommended limit (45 mg 1-1) especially for those samples occurring in rural areas. Almost 65% of samples from dug wells showed concentrations greater than 45 mg 1-1.

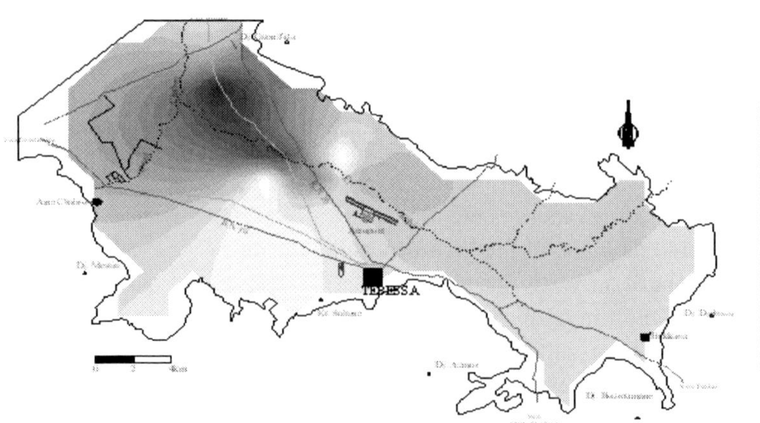

Figure 3.2. Nitrate distribution in groundwater of the study area. (mg l^{-1}).

The presence of high nitrate concentrations in the phreatic waters is not only the direct consequence of a massive usage of artificial fertilizers, but is also a consequence of contamination by domestic septic tanks. The latter are unfortunately favoured by the absence of a sanitation network in the whole region where more than 800 septic tanks are in use. This fact contributes to the nitrification of groundwaters according to the process.

Nitrate concentrations vary greatly from18 mg l-1 at P11 to 80 mg l-1 P18. The highest concentrations were found in the middle and the west part of the studied area.

3.3. Anthropogenic Inputs

Lithology and human activities controlled the hydrogeochemical characteristics of groundwater. Variation in TDS in groundwater may be related to land use and also to pollution (Rouabhia 2009). Cl^-, SO_4^{2-}, NO_3^- and Na^+ ions are mostly derived from agricultural practices and use of wastewater in irrigation, animal wastes and industrial and municipal sewage (Rouabhia 2010).

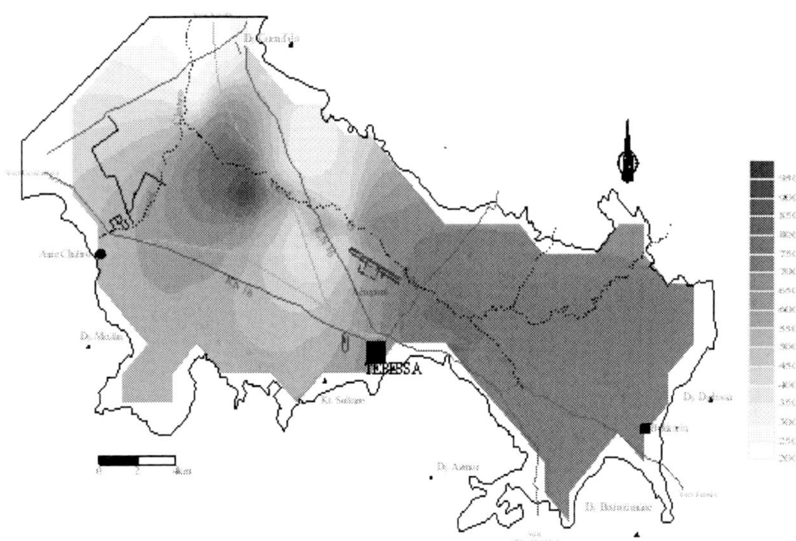

Figure 3.3. Sulfate distribution in groundwater of the study area. (mg l^{-1}).

Correlation of these ions with TDS can be used to indicate the influence of human activities on the water chemistry. The relationship between various anions and cations with TDS is illustrated in Figure4. The Na+ concentrations show an increasing trend with increasing TDS (Figure4.1) and in addition to the weathering of silicate minerals it may be related to the anthropogenic sources such as sewage, household waste, engineering work effluents, deicing road salt, etc. Increase of Ca^{2+} concentrations with increasing TDS (Figure 4.2, 4.3) supports the anthropogenic input mainly domestic and industrial waste (Rouabhia 2009). Cl^- and NO_3^- show a good trend of increasing concentration with increasing TDS (Figure 4.4, e), suggesting same source and can be used as pollution indicators for anthropogenic input (Rouabhia 2008). The SO_4^{2-} appears to show a good trend of increasing concentration with increasing TDS.

In presence of geological source the other possible source of SO_4^{2-}in study area is mainly use of wastewater in irrigation, industrial effluents and domestic sewage. A poor correlation between NO_3^- and SO_4^{2-} is observed suggesting different source of both ions as agricultural activity observed in the study area (Figure 5.2).

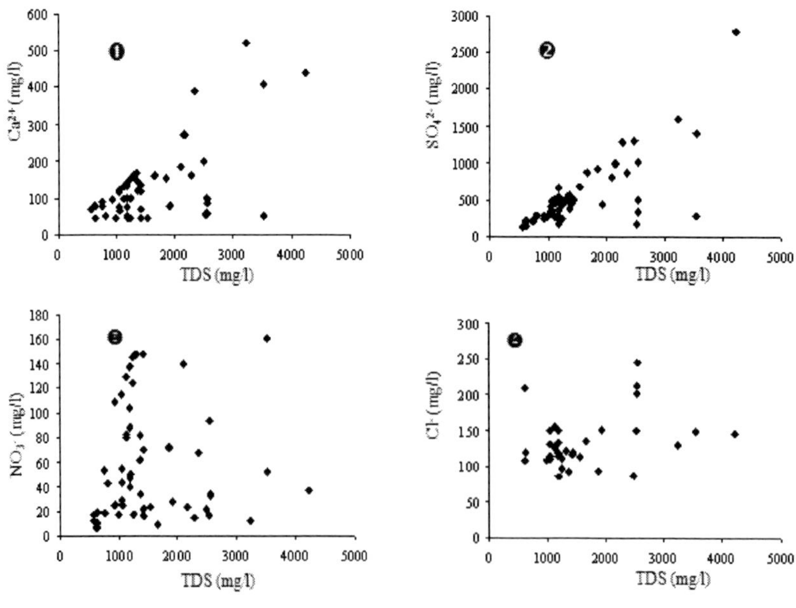

Figure 4. Relationship between TDS and cations/anions in groundwater of the study area

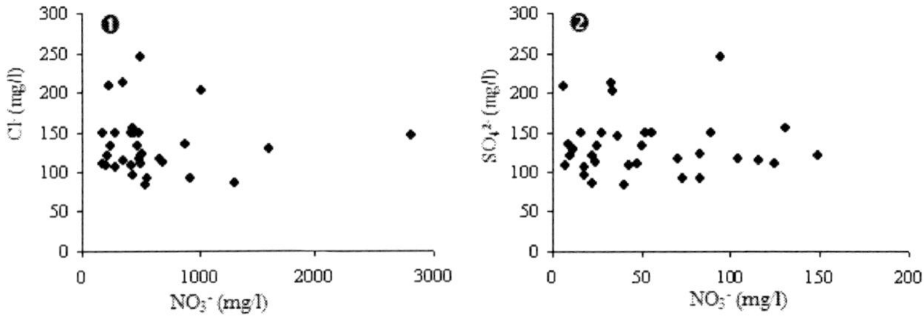

Figure 5. Relationship between **a** NO_3^- and Cl^- and **b** SO_4^{2-} and NO_3^-.

Higher correlation between Cl^- and NO_3^- above 0.35, indicate common source and input from anthropogenic activities. In the study area positive correlation between Cl^- and NO_3^- ($r = 0.54$) (Figure 5.1) have been observed, suggesting same source for these ions.

4. Conclusion

Present study adds to our understanding of the impact of urbanization on groundwater resources. Chemical properties of groundwater are controlled both by natural geochemical processes and anthropogenic activities. Our studies suggest that Na^+ is mostly derived from both natural as well as anthropogenic sources. The variation of TDS and its correlation with other ions supports the contribution of anthropogenic inputs to the groundwater contamination. The major pollutant in the groundwater of study area is NO_3^-. High concentration of NO_3^- is due use of wastewater in irrigation. Majority of the samples have high NO_3^- concentration and become unfit for human consumption, particularly infants. Majority of the samples are good for irrigation in almost all type of soils with little danger of exchangeable sodium as they fall in very high salinity–low sodium field.

The study reveals that the central part of the study area is more prone to groundwater pollution owing to the generation of bulk sewage and solid waste coupled with improper disposal of sewage and urban solid waste. In the study area, pheratic zone is more polluted than the deeper zone and hence preventive steps should be taken up to protect them.

ACKNOWLEDGMENTS

This work has been realized through the framework of CNPRU project G02920100015 under the thematic: *Mécanismes d'acquisition de la minéralisation des eaux souterraines dans la plaine d'effondrement de Tébessa N.E Algérie*. We would like to thank Prof Younsi Abdelkader, University of El Djadida (Morroco).

REFERENCES

[1] Blés JL, Fleury JJ (1970) Carte géologique de l'Algérie au 1/50000: feuille n°178, Morsott, avec notice explicative détaillée. Service de cartes Géologique et Sonatrach, Division d'hydrocarbure. Direction des explorations, Alger, Algérie.

[2] Djabri L (1987) Contribution to the hydrogeological study of the subsidence plain of Tebessa NE Algeria. Attempt of modelling. Doctorate Thesis, University of Franche Comté, France.

[3] Edmunds WM, Shand P, Guendouz A, Moulla AS, Mamou A, Zouari K (1997) Recharge characteristics and groundwater quality of the Grand Erg Oriental basin, final report. EC (Avicenne), Contract CT93AVIOOI5, BGS Tech Rep WD/97/46R, *Hydrogeology Ser.*

[4] Girard P, Hillaire M, Oga MS (1997) Determining the recharge mode of Sahelian aquifers using water isotopes. J Hydrol 197:189–202.

[5] Parkhurst DL, Apello CAJ (1999) User guide to PHREEQC (version 2)-a computer program for speciation, batch reaction, one dimensional transport, and inverse geochemical Calculations: *U.S. Geological Survey Water Resources Investigations Report*: 99–4259, 312 p.

[6] Plummer LN, Jones BF, Trusedall AH (1976) WATEQ-a Fortran IV version of WATEQA computer program for calculating chemical equilibrium of natural waters. U.S Geol-Surv. Water Res, Washington DC, vol 76, pp 13–61 (Revised 1978, 1984(

[7] Rouabhia A, Baali F, Kherici N, Djabri L (2004) Vulnérabilité et risque de pollution des eaux souterraines de la nappe des sables miocènes de la plaine d'El MA EL Abiod (Algérie). *Rev Sécheresse* 15:347–352

[8] Rouabhia A (2006) Vulnérabilité et risque de pollution des eaux souterraines de la nappe des sables mioce`nes de la plaine d'El MA EL Abiod (Algérie). Doctorat Thesis, University of Annaba, Algeria.

[9] Rouabhia A., Ch. Fehdi, F. Baali, L. Djabri, R. Rouabhi (2008) Impact of human activities on quality and geochemistry of groundwater in the Merdja area, Tebessa, Alegria. *Environ Geol.* DOI:10.1007/s00254-008-1225-0 .

[10] Rouabhia A. F. Baali, Ch. Fehdi, N. Kherici, L. Djabri (2008): Hydrochemical and isotopic investigation of a sandstone aquifer groundwater in a semi arid region, El Ma El Abiod, Algeria. Environ Geol. DOI:10.1007/s00254-008-1451-5.

[11] Rouabhia A. Baali F. Kherici N. Djabri L. (2004) Vulnérabilité et risque de pollution des eaux s/t de la nappe des sables miocènes d' El Ma El Abiod (Algérie). Sécheresse n°4, vol. 15 (France.(

[12] Fehdi Ch. Rouabhia, A. Baali, F. Boudoukha A (2009): The hydrogeochemical characterization of Morsott-El Aouinet aquifer, Northeastern Algeria. *Environ Geol.* DOI: 10.1007/s00254-008-1667-4.

[13] A. Rouabhia, F. Baali, Ch. Fehdi (2009) «Impact of agricultural activity and lithology on groundwater quality in the Merdja area, Tebessa, Algeria». *Arabian journal of Geosciences*, By Springer. DOI: 10.1007/s12517-009-0087-4.

[14] Younsi A. (2001). Méthodologie de mise en évidence des mécanismes de salures des eaux souterraines côtières en zone semi-aride irriguée. Chaouia côtière, Maroc. Thèse de doctorat d'état. Univ. Chouaib Doukkali. El Jadida Maroc, 18-30.

[15] Sawyer C, McCarthy P (1967) Chemical and sanitary engineering, 2nd edn. McGraw-Hill, New York.

[16] Todd D (1980) Ground water hydrology, 2nd edn. Wiley, New York Vila JM (1980) La chaıˆne alpine de l'Algérie orientale et des confins Algéro-Tunisiens. Thèse de Doctorat e`s sciences, Université Pierre et Marie curie, Paris VI.

[17] WHO (1998) Guidelines for drinking water quality. Addendum to vol 2, 2nd edn. Health criteria and other supporting information (WHO/EOS/98.1), World Health Organization.

In: Aquifers: Types, Impacts and Conservation ISBN: 978-1-61942-091-5
Editors: Z. Ouakili, et.al, pp. 149-167

Chapter 8

NUMERICAL STUDY OF AQUIFER THERMAL ENERGY STORAGE SYSTEM INFLUENCED BY REGIONAL GROUNDWATER FLOW

Kun Sang Lee[1]
Department of Environmental and Energy Systems Engineering,
Kyonggi University, Suwon, Kyonggi, South Korea

ABSTRACT

This paper presents numerical investigations and thermohydraulic evaluation of two-well models of aquifer thermal energy storage (ATES) system operating under continuous flow regime. A three-dimensional numerical model for groundwater flow and heat transport is used to analyze the performance of thermal energy storage in the aquifer. Thisstudy emphasizes the effects of regional groundwater flow on the performance of the ATES system under various operation scenarios consisting of different parameters including flow conditions of injecting and producing water (temperature and flow rate), regional groundwater velocity and properties (hydraulic gradient), and aquifer characteristics (heat loss and size). Performances of ATES were compared in terms of temperature of extracted water and temperature field in the aquifer. The calculated temperature at the producing well varies within a certain range through the year and fluctuating quarterly a year, reflecting the

[1]Corresponding author: (Phone) 82-31-249-9738, (Fax) 82-31-244-6300, (E-mail) kslee@kyonggi.ac.kr.

seasonaltemperature conditions of injecting water. The pressure gradient across the system, which determines the direction and velocity of regional groundwater flow, has a substantial influence on the aquifer thermal storage. Water flow opposite to the direction of regional groundwater flow leads to an effective ATES because of smaller loss and less fluctuation in extracted thermal energy. Injection/production rate and geometrical size of the aquifer used in the model also impact the predicted temperature distribution at each stage and the recovery water temperature. But heat loss with the adjacent confining strata hasa minimal effect on the performance of ATES systems. The hydrogeological-thermal simulation is shown to be an integral part in the prediction of performance for a process as complicated as ATES systems.

Keywords: Aquifer thermal energy storage, Hydrogeological-thermal simulation,Groundwater flow, Continuous operation

INTRODUCTION

Energy is a key element in interactions between nature and society and is considered a key input for economic development. As the demand for energy increases, any work to enhance energy conservation is crucial. Thermal energy storage (TES) system applications around the world have been known to provide economical and environmental solutions to the energy problems [1]. TES system has a huge potential to improve energy efficiency and reduce environmental impacts for energy systems by matching energy supply and demand. Thermal energy can be stored in different ways given by the thermodynamics of the storage process. Underground thermal energy storage (UTES) is mostly used for seasonal heat/cold storage [2]. There are several concepts as to how the underground can be used for UTES depending on geological, hydrogeologicaland other site conditions. The most frequently used storage technology among various types of UTES systems is aquifer thermal energy storage (ATES) system using groundwater to carry the thermal energy into and out of an aquifer to utilize the low-temperature geothermal resource in the aquifer [3, 4]. Aquifer thermal energy storage, which is similar to the groundwater geothermal system under direct uses, involves storage and provides for both heating and cooling on a seasonal basis. The high heat transfer capacity and low operational cost make ATES usually the cheapest alternative, if the aquifer is hydrogeologically suited. ATES systems hold great promise for energy storage and have been used successfully around the world for the seasonal storage of heat and cold energy for the purpose of heating and cooling buildings [5-8].

Estimation of heat recovery rate and temperature distribution in the aquifer is required in ATES projects before the system is installed. At present, the use of computer modeling constitutes an integral part in the prediction and evaluation of geothermal performance [9, 10]. In carrying out ATES development projects, a numerical modeling based on coupled mass and energy transport theory has to be conducted on the behavior of local subsurface geothermal system to evaluate and optimize a project design.

A number of researchers have highlighted the important role of numerical modeling in the analysis of ATES systems. Molson *et al.* [11] used a three-dimensional finite element model to simulate groundwater flow and energy transport in an unconfined aquifer. Probert *et al.* [5] presented the thermodynamic evaluations of ATES projects and listed key aquifer properties and design parameters. Based on an elementary ATES model, Rosen [12] performed second-law analysis to thermal energy storage systems to assess overall system performance. Chevalier and Banton [13] applied the random walk method of resolution to the study of energy transfer phenomena in ATES using a single injection well. Tenma *et al.* [14] and Lee and Jeong [15] carried out a more realistic two-well model study to examine an underground design of TES system.

Any ATES design and application will require the sensitivity of parameters affecting on the long-time performance of aquifer being target to use. The most important properties are geometry (surface area), stratigraphy (properties of different layers), groundwater gradient (natural flow direction and velocity), and operation condition (flow rates of injection and extraction) [16]. Previous studies, however, derived conclusions based on the simulation results for ATES systems without considering the effects of regional groundwater flow. It has been well known that the groundwater flow would have a significant impact on the heat transfer [17, 18]. To overcome a limited applicability of previous researches, more comprehensive study should be established for the evaluation of the ATES systems. The present work extends previously reported researches to ATES systems under groundwater flow. The main design considerations involve heat losses, injection/production rates and configuration of well-aquifer system.

Coupled hydrogeological-thermal simulations were undertaken in order to predict thermal behavior of aquifer and recovery temperatures from the aquifer. Analyses for a two-well system in a confined aquifer were performed in order to determine howgroundwater flow in response to hydraulic gradient affects results of aquifer thermal energy storagesimulations. The aim of the evaluation is to determine whether groundwater flow needs to be taken into consideration to make reliable predictions about future recovery temperatures and temperature distributions in the aquifer given the planned injection/production temperatures and rates.

MATHEMATICAL THEORY

To calculate temperatures of the aquifer at different locations, theoretical principles of water flow and heat transfer phenomena were explained. There are a number of simulation models available on the market. Among the most advanced simulators, a general simulatornamed UTCHEM has proved to be particularly useful formodeling multiphasetransport processes under nonisothermal condition [19]. UTCHEM has been extensivelyverified bycomparing to analytical solutions and experimentalmeasurements for its abilityto predict the flow of fluids through the aquifer.

The continuity of mass for water in association with Darcy's law is expressed as

$$\frac{\partial}{\partial t}(n\rho_w) + \nabla \cdot (\rho_w \mathbf{u}) = R_w \tag{1}$$

where

n: porosity [dimensionless]
ρ_w: density of water [ML^{-3}]
u: Darcy flux [Lt^{-1}]
t: time [t]
R_w: source term [$ML^{-3}t^{-1}$]
The water flux from Darcy's law is
The water flux from Darcy's law is

$$\mathbf{u} = -\frac{\mathbf{k}}{\mu_w}(\nabla p - \gamma_w \nabla z) \tag{2}$$

where k is the permeability tensor; μw the dynamic viscosity [ML−1t−1]; p the pressure [ML−1t−2]; z the vertical depth [L], and γw the specific gravity (= ρwg) [ML−2t−2].

The energy balance equation is derived by assuming that energy is a function of temperature only and energy flux in the aquifer occurs by convection and conduction only. The resulting general heat balance equation describing nonisothermal groundwater flow in a saturated porous medium can be formulated as follows;

$$\frac{\partial T}{\partial t}\left[(1-n)\rho_r C_{vr} + n\rho_w C_{vw}\right] + \nabla\cdot\left(\rho_w C_{vw}\mathbf{u}T - \lambda_T \nabla T\right) = q_H - Q_L \tag{3}$$

where,

T: aquifer temperature [T]

C_{vr}, C_{vw}: rock and water heat capacity at constant volume $[ML^{-1}t^{-2}T^{-1}]$

λ_T: thermal conductivity of aquifer $[MLt^{-3}T^{-1}]$

q_H: enthalpy source per unit bulk volume $[ML^{-2}t^{-2}]$

Q_L: heat loss with overburden and underburden formations $[ML^{-2}t^{-2}]$

Thermal dependence of density, viscosity, thermal conductivity, and heat capacity is not taken into consideration because these parameters vary little in the considered temperature range.

The heat transfer with over and underlying low-permeability layers, is assumed to be due solely to thermal diffusion and the heat equation is simplified to

$$Q_L = \nabla\cdot\left(\lambda_{Te}\nabla T\right) \tag{4}$$

where λ_{Te} is the thermal conductivity of overburden or underburden rock $[MLt^{-3}T^{-1}]$. The heat transfer equation (3), which results from the principle of energy conservation, is coupled with the flow equation from Darcy's law (2) and the continuity equation (1).

MODELING

An understanding of the thermohydraulic processes in the aquifer is necessary for a proper design of an ATES system under given thermohydraulic conditions. Simulations on three-dimensional Darcy's flow in the aquifer were carried out with a finite-difference model for ground-water flow and heat transport. The numerical model includes the effects of hydraulic anisotropy, forced and natural convections and conduction, and heat losswith the adjacent confining strata. In order to decide the sustainability of the aquifer for energy storage application, temperatures of producing water is estimated over a ten-year period. To reduce the numerical dispersion associated with solution of Eq. (1) in finite-difference form, total-variation-diminishing (TVD) third order method was used in applying the block-centered finite-difference approximation of spatial derivative [20].

In order to estimate theparameters of an underground system, a general ATESusing the open system of two-wells is considered. As shown in Figure 1, two wells are situated at the centers of squares consisting of a 2×1 rectangular field, one to extract the groundwater and the other to reinject it into the same aquifer it was produced from. The transfer of thermal energy is realized by pumping water into the injection well at a constant flow rate$Q$ at the surface temperatureT_{inj} and extracting it at the same flow from the neighboring production well. In a large reservoir with repeated patterns, the flow is symmetric around each pair of injection-injection wells confined to the pattern. Constant pressure is specified to left and right boundaries to induce regional groundwater flow occurred in response to hydraulic gradient and the other boundaries are closed to flow. All the outer boundaries are represented as adiabatic to simulate symmetry in an array. The heat transfer in the ATES was inherently coupled with heat conduction and advection by flow of injecting water and regional groundwater.

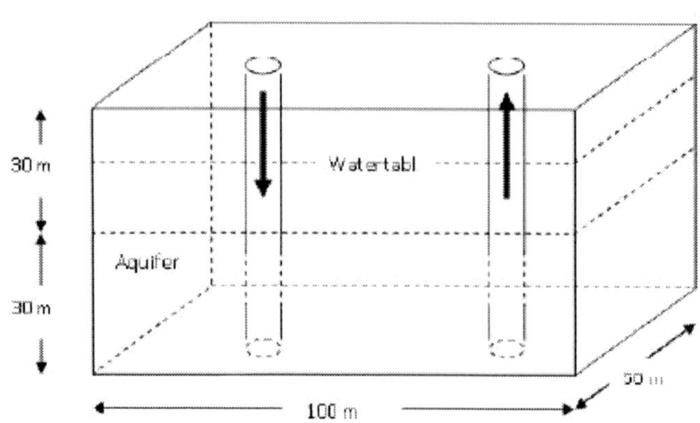

Figure 1.Schematic representation of simulation domain of base case.

Geometry of the model is similar to that of the two-well model suggested by Tenma *et al.* [14]. The size ofthis two-well model is 100 m × 50 m × 30 m for a base case. This model is a unit ofthe system and composed of two wells partially-penetrating a confined aquifer.The grid is divided into 22and 11 grid blocks in *x* and *y*directions,and 10 gridblocksinaverticaldirection. Grid block sizes range 2-10 m in horizontal direction and set to be constant in vertical direction. Welldepthwas set at the relativelyshallowvalueof for thetwo-well system. Water level in the model was set at −15 m. Top and bottom levels of the confined aquifer were set at −30 m and −60 m, respectively. The model described here comprises a grid unit systemwith two pipes of 0.2 m radiusthat partially-penetrate a confined aquifer.Injection and production pipes arepenetrating the ground at the intervals of 21-27 m and 3-9 m from the top of the aquifer, respectively.

Determination of the potential of a specific confined aquifer as an effective thermal energy storage medium requires thorough knowledge of thermodynamic and hydraulic properties of the aquifer and its confining layers and fluid properties. Parameters include porosity and permeability of the storage aquifer and thermal conductivities and heat capacities of the aquifer matrix, native ground water,and confining layers.

Table 1. Hydrogeological and thermal properties of aquifer and water

aquifer	porosity (n)	0.40
	permeability (k)	1,013 md
	compressibility of formation (β_r)	2.96×10^{-6} kPa^{-1}
	density of rock (ρ_r)	2.65 g/cm^3
	thermal conductivity of rock (λ_T)	200 kJ/day·m·K
	thermal conductivity of overburden/underburden rock (λ_{Te})	249.2 kJ/day·m·K
	heat capacity of rock (C_{vr})	0.8864 kJ/kg·K
water	viscosity (μ_w)	1.1404 cp
	compressibility (β_w)	4.4×10^{-7} kPa^{-1}
	density (ρ_w)	1 g/cm^3
	heat capacity (C_{vw})	4.184 kJ/kg·K

As presented in Table 1, the aquifer and water have constant thermal properties and assumed to be slightly compressible. The aquifer is assumed to be continuous and its thermal conductivity and volumetric heat capacity are considered to be a function of porosity and the thermal characteristics of water and soil matrix [21].

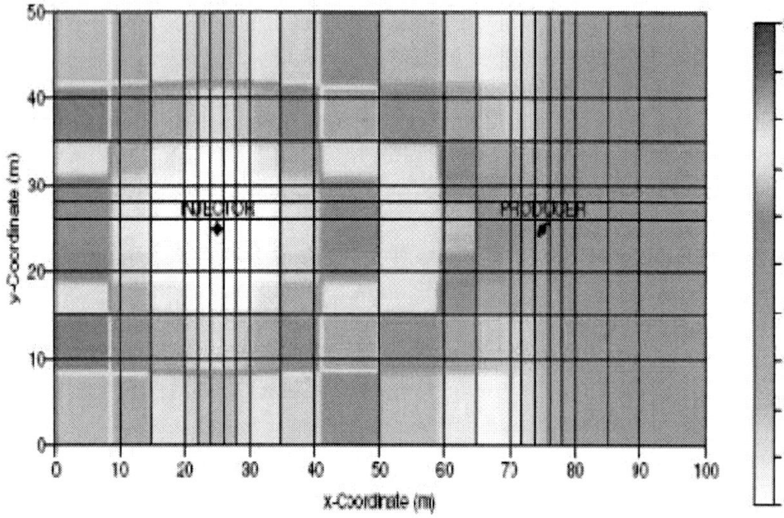

(a)no groundwater flow.

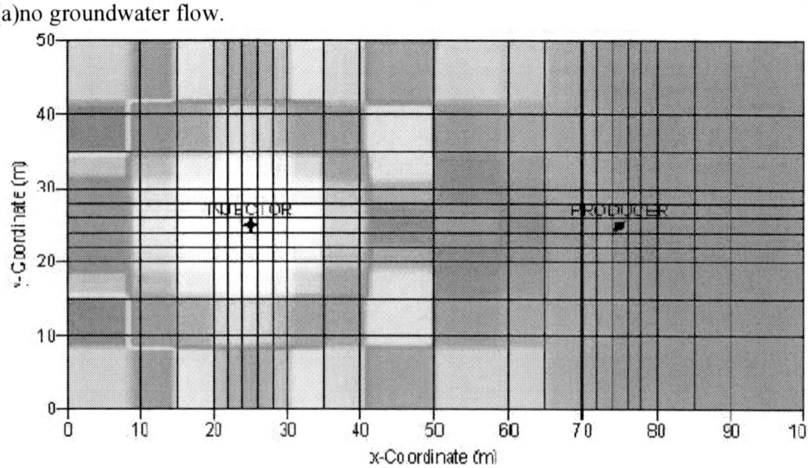

(b) groundwater flow from right to left ($\Delta p = -20$ kPa).

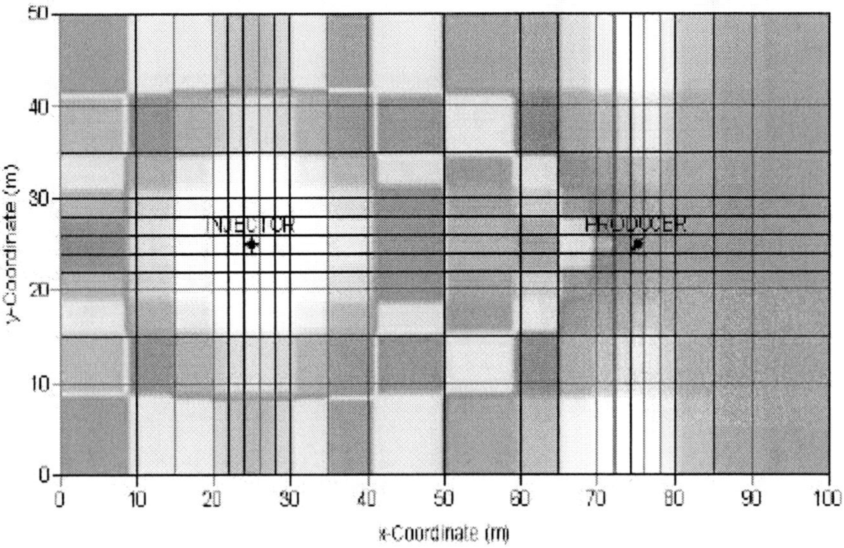

(c) groundwater flow from left to right (Δp = 20 kPa).

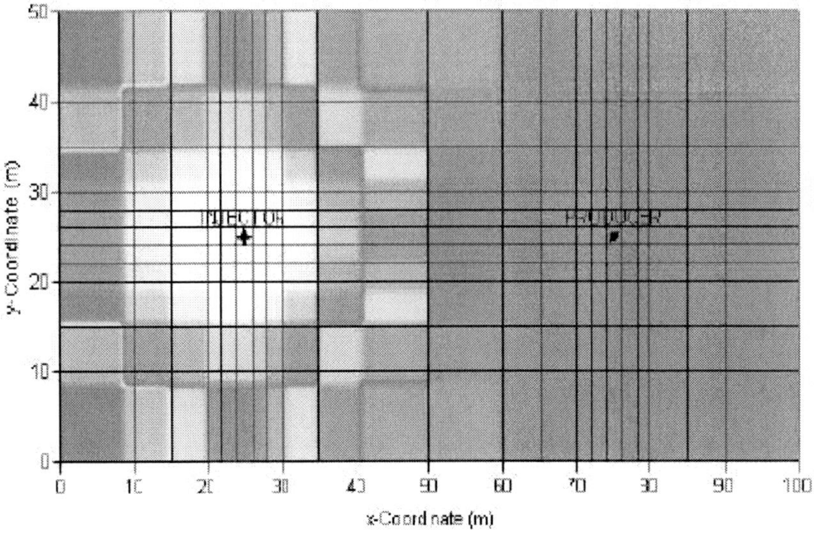

(d) groundwater flow from right to left (Δp = –40 kPa).

Figure 2. (Continued)

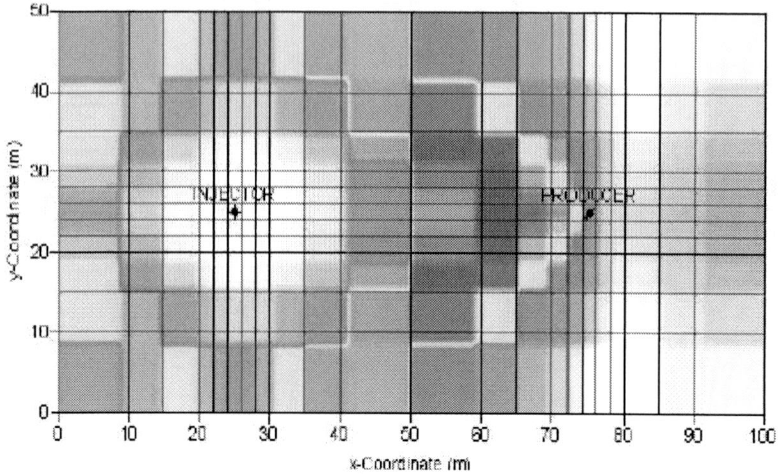

(e) groundwater flow from left to right (Δp = 40 kPa).

Figure 2. Temperature distribution [°C] obtained with different groundwater flow after cold water injection.

RESULTSAND DISCUSSION

To estimate the characteristics of the system, the two-well model was run for continuous flow regime. Heat transfer within the aquifer is simulated by specifying constant temperatureT_{inj} at the injection well and with the aquifer temperature initialized at T_i. The initial temperature of the aquifer is assumed to be constant 17.5°C over the entire aquifer and confining layers. The pressure difference between left (p_L) and right (p_R) boundaries ranges from –40 kPa to 40 kPa, and derives regional ground flow from left to right for $\Delta p > 0$ or right to left for $\Delta p < 0$. The model is run for 10 years to provide an adequate long-term assessment of thermal storage.

The ATES is operated under the continuous flow regime, where water is pumped from one well equipped with a pump and injected through a second well. A complete energy storage cycle is composed of four periodsper year to simulate the seasonal conditions. Each cycle is symmetrical for identical injection (Q> 0) and production (Q< 0) rates and duration. The thermal field and the temperature of produced water, T_{prod}, are calculated from the numerical solutions of Equation (3) at constant time interval.

Groundwater Flow

Accounting for the seasonal changes in surface temperature, theinjected water temperaturesweretaken as 5°C, 15°C, 25°C, and 15°C over a three-month period, respectively. The size ofthis two-well model is 100 m × 50 m × 30 m for base case. The rates of injection or production of 100 m^3/day correspond to 0.2688 pore volume of the aquifer for three months. To evaluate the sensitivity of groundwater flow to the performance of BTES, recovery temperature values were calculated with the boundary conditions of $\Delta p = 0$ (no regional groundwater flow), 20, and 40 kPa corresponding to Darcy fluxes of 0, 0.018, and 0.035 m/day, respectively.

Figures 2(a) to (e) show temperaturedistribution from the numerical calculations under the boundary conditions of $\Delta p = 0$, –20, –40 (regional groundwater flow from right to left), 20, and 40 kPa (regional groundwater flow from right to left) after 90 days of cold water injection.

From the figures, the effect of groundwater flow is clear and it influences the thermal flow significantly. When the direction of regional groundwater flow is same with that of injected water, i.e. $\Delta p > 0$, the thermal front approaches the producing well much closer due to increased thermal convection, and thermally impacted area is much larger. For the case of repeated patterns in larger area, the thermal front can be somewhat overlapping each other. This will be a disadvantageous condition for the recovery temperature.

Simulation studies on the heat transfer in the ATES have been carried out with and without the presence of regional groundwater flow. The temperatures of produced water are shown in Figure 3. Balancesbetween injecting and producing thermal energy and small variationin temperatures are the most desirable case, representing little fluctuation in extracted thermal energy and sustainable use of ground as a BTES system without gradual heating or cooling.As shown in Figure 3, the temperaturegradually decreases in all the cases. As thetemperature of the circulated fluid changes by 25-5°C, the temperature eventually becomes 15°C, the average temperature of injected water. Due to energy imbalance between the initial temperature of 17.5°C and average injecting temperature of 15°C, the aquifer undergoes a gradual cooling process. As the pressure difference becomes larger, the cooling process is faster.

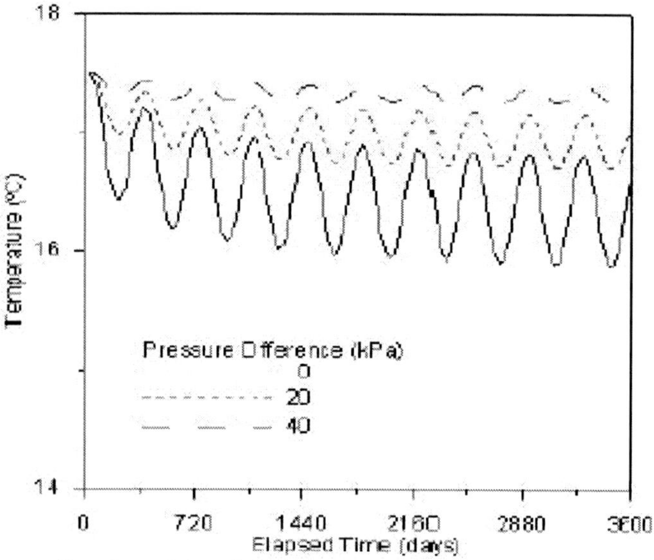

(a)groundwater flow from right to left.

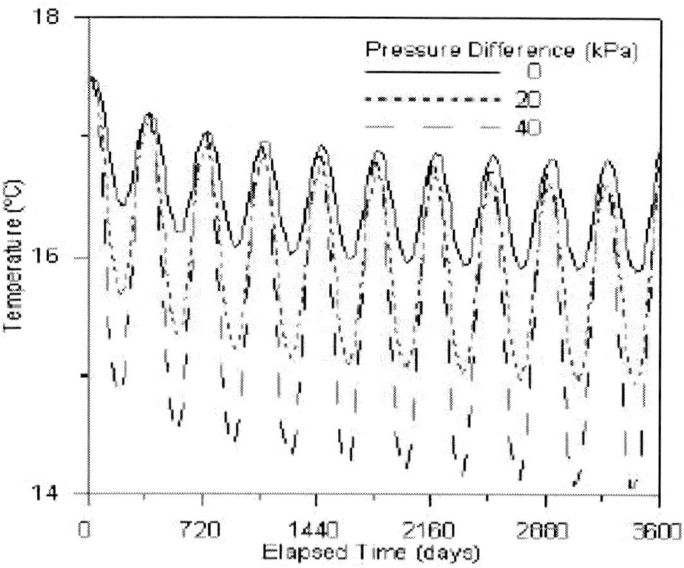

(b) groundwater flow from left to right.

Figure 3. Temperature of produced water obtained from simulations with different groundwater flow condition for q =100 m3/day.

Reflecting the changes in temperature of the injecting water, the temperature of recovered water fluctuates with a quarterly year period. The temperatures at the producing well were constant within the range of 0.94°C to 2.60°Cfor the case of left to right groundwater flow and 0.16°C to 0.94°C for the case of right to left groundwater flow through the ten-year period. Differences in temperature variation among cases are quite significant, more than 2.7 or 5.8 times depending on the pressure difference. The range of variation of Δp = 40 kPa is the highest and 16 times higher than that of Δp = -40 kPa which is the smallest.

As the net change of thermal energy is small, the groundwater flow conditions under large pressure gradient and opposite direction to the injected water are more promising.

Heat Loss

Heat transfer from/to overburden and underburden formations is considered as an important factor that controls the temperature of produced water. In the present work, results from the base case were compared with those from a case in which heat loss is not included. These formations are assumed to have the same thermal properties with aquifer, as stated earlier.

Figure 4 shows these results at different times with Δp of 0, ±20, and ±40 kPa. The larger Δp, the smaller difference between results is obtained. Differences in temperature variation among cases range from 5.0% to 14.2% depending on the pressure difference. With groundwater flow from right to left, the inclusion of heat loss effect makes relatively small difference. Contrarily, the temperature at the production well shows considerable difference under the groundwater flow from left to right. When heat loss is included during the simulation, the temperature remains higher and temperature variation is smaller due to overall heat transfer from overburden and underburden formations. The heat transfer results from higher temperature of confining layers compared with average temperature of injecting water. The results indicate that conductive heat loss with the surrounding rock may be an important process causing different cycle of temperature variations in the production well, especially for the groundwater flow under large pressure gradient and same direction to the injected water.

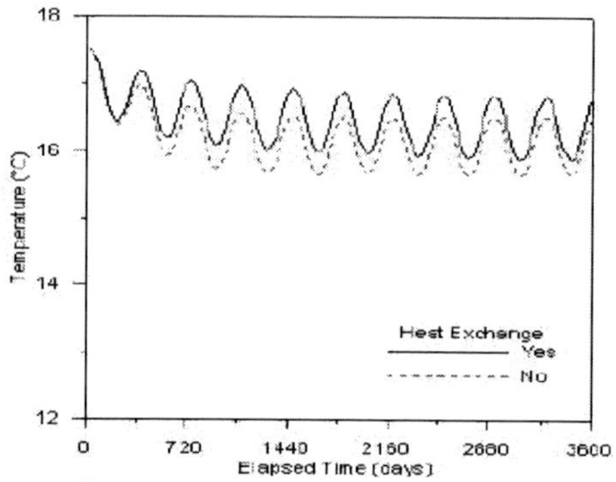

(a) no regional groundwater flow.

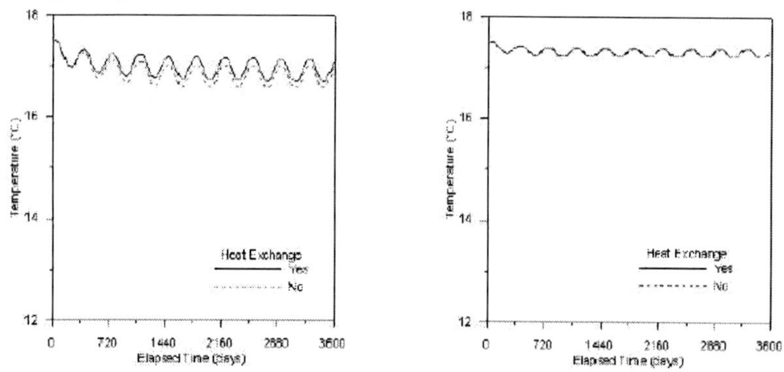

b) groundwater flow from right to left (Δp = 20 and 40 kPa.)

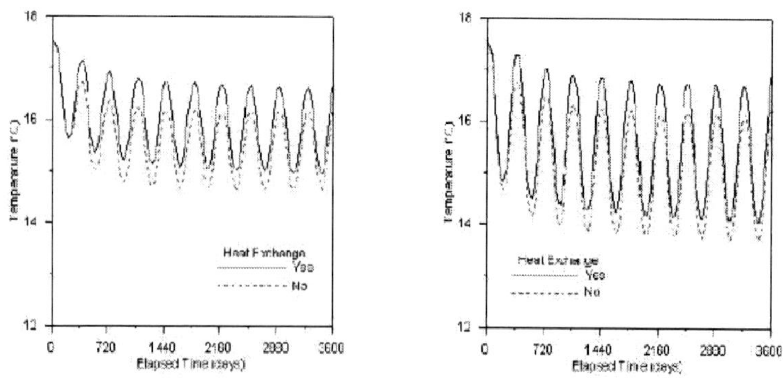

(c) groundwater flow from left to right (Δp = 20 and 40 kPa).

Figure 4. Temperature of produced water obtained from simulations with different heat loss conditions for q =100 m3/day.

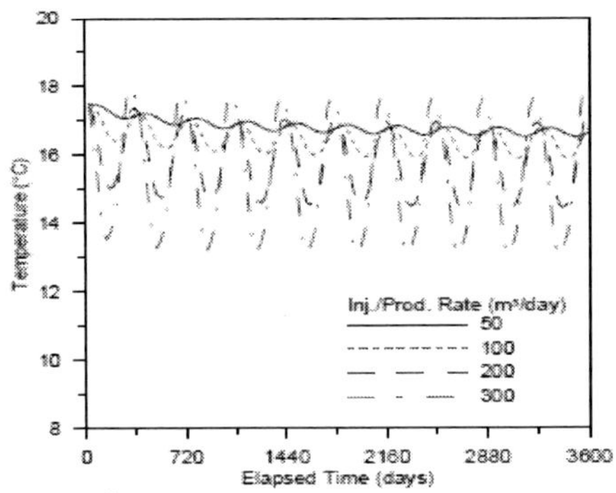

(a) no regional groundwater flow.

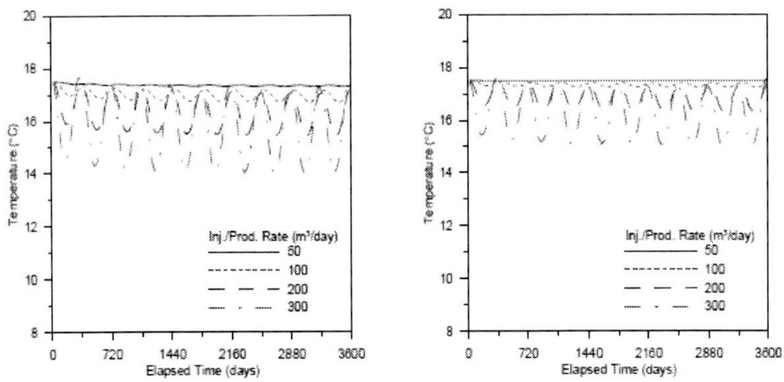

(b) groundwater flow from right to left (Δp = 20 and 40 kPa).

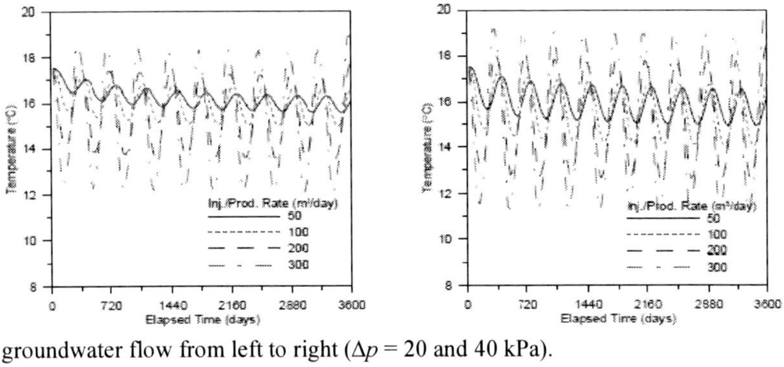

(c) groundwater flow from left to right (Δp = 20 and 40 kPa).

Figure 5. Temperature of produced water obtained from simulations with different injection/production flow rates.

Flow Rate

The aim of this simulation is to evaluate the recovery of thermal energy from the aquifer given injection or production flow rates. The calculations were performed for well and aquifer configuration which is the same as the base case. However, in this simulation, the flow rates are changed to 50, 100, 200, and 300 m^3/day.

The profiles in Figure 5 shows increasing range in temperature of recovered water with increasing flow rate in accordance with the gradual drop over time.

By increasing the flow rate from 50 m^3/day to 300 m^3/day, a significant increase in the variation of temperature is obtained. Increasing the flow rate six times results in the increase of temperature variation from 0.01°C to 2.61°C for $\Delta p = -40$ kPa and from 2.57°C to 8.56°C for $\Delta p = 40$ kPa.

The result suggests that, everything else being the same, the use of less flow rate opposite to the direction of regional groundwater flow is a promising flow condition due to small net change in thermal energy at the production well. However, decreasing the flow rate considerably less than an appropriate value would not be effective because the resulting decrease in the amount of available thermal energy would be a problem.

Aquifer Size

One objective with the numerical simulations is to find the well configuration to make the energy storage as dense as possible. In attempts to investigate effects of interwell distance, pumping and injection of groundwater were simulated for four cases considered, 80 m × 40 m, 100 m × 50 m, 150 m × 75 m, 200 m × 100 m. The injection and production wells are located 40 m, 50 m, 75 m, and 100 m apart, respectively.

Figure 6 illustrates a longer well distance reduced the variations in thermal storage substantially. Larger variations for shorter well-to-well distance result from the fact that the thermal front of injected water approaches the production well within each operation period. Therefore, the region near a producing well is considerably affected by injected water. This observation emphasizes the importance of ensuring that adequate distance between wells is used, taking into account the thermal and hydraulic transport of injected water.

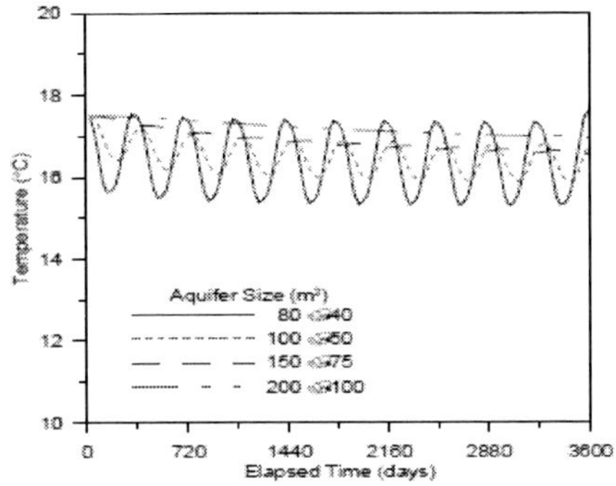

(a)no regional groundwater flow.

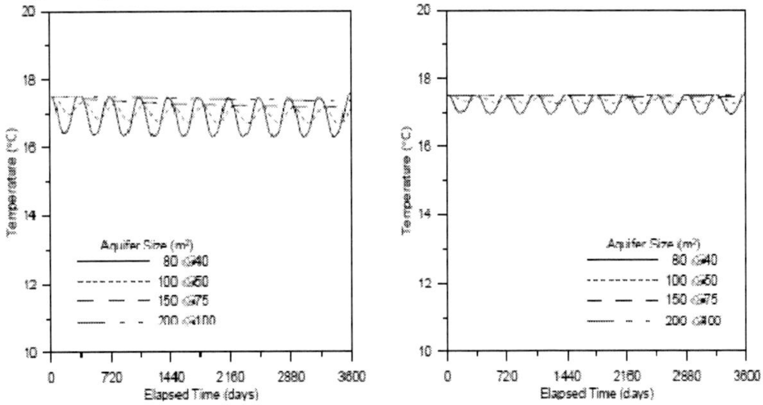

(b) groundwater flow from right to left (Δp = 20 and 40 kPa).

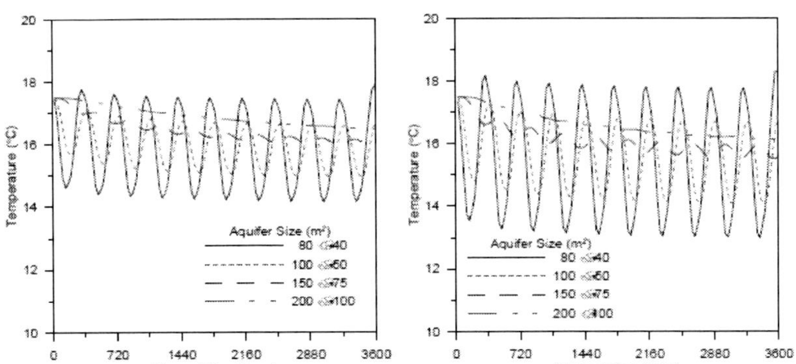

(c) groundwater flow from left to right (Δp = 20 and 40 kPa).

Figure 6. Temperature of produced water obtained from simulations with different aquifer size for q = 100 m3/day.

CONCLUSION

Numerical calculations were carried out to estimate long-time thermal behavior of ATES system with two wells under continuous operation methods. The effects of various geometrical and operational parameters on computed values of aquifer thermal behavior and final producing temperatureunder regional groundwater flow were studied for a 10-year continuous injection and withdrawal.The hypothetical simulations indicate that the model of the two-well system will be a valuable tool in determining the sensitivity of various parameters.

The heat transfer in the ATES was affected by a number of factors such as flow conditions of injecting and producing water (temperature and flow rate), groundwater velocity and properties (hydraulic gradient), and aquifer characteristics (heat loss and size). The thermal behavior of the storage system is shown to depend on the direction and velocity of the groundwater flow, which is determined by regional pressure gradient. Low flow rate opposite to the direction of regional groundwater flow and large aquifer size are recommended as an effective ATES because of small loss and little fluctuation in extracted thermal energy. When the direction of regional groundwater flow is same with that of injected water, the thermal front approaches the producing well much closer due to increased thermal convection and results in large variations in the temperature of recovered water. Difference in temperature variation among cases depends on the pressure difference.

REFERENCES

[1] Paksoy, HO; Gurbuz, Z; Turgut, B; Dikici, D; Evliya, H. *Renewable Energy,* 2004, 29, 1991-1996.

[2] Nielsen, K.Thermal Energy Storage, A State-of-the-Art; NTNU: Trondheim, Norway 2003.

[3] Sanner, B. *GHC Bulletin*, 2001, 22, 19-25.

[4] Rafferty, K. *Groundwater*, 2003, 41, 408-410.

[5] Probert, T; Hellsröm, G; Glaesson, J. *Proc. Int. Symp. on Aquifer Thermal Energy Storage,* 1994, 73-81.

[6] Paksoy, HO; Andersson, O; Abaci, H; Evliya, H; Turgut, B. *Renewable Energy*, 2000, 19, 177-122.

[7] Allen, DM; Ghomshei, MM; Sadler-Brown, TL; Dakin, A; Holtz, D. *Proc. World Geothermal Congress*, 2000, 55-58.
[8] Schmidt, T; Mangold, D; Muller-Steinhagen, H. Proc. *ISES Solar World Congress,*2003, 1-7.
[9] Breger, DB; Hubbell, JE; Hasnaoui, HE; Sunderland, *JE. Solar Energy,* 1996, 56, 493-503.
[10] O'Sullivan, MJ; Pruess, K; Lippmann, *MJ. Geothermics*, 2001, 30, 395-429.
[11] Molson, JW; Frind, EO; Palmer, CD. *Water Resources Research*, 1992, 28, 2857-2867.
[12] Rosen, MA. *Energy*, 1999, 24, 167-182.
[13] Chavalier, S; Banton, B. *J. Hydrology*, 1999, 222, 129-139.
[14] Tenma, N; Yasukawa, K; Zyvoloski, G. *Geothermics*, 2003, 32, 603-607.
[15] Lee, KS; Jeong, SJ. *Int. J. Green Energy*, 2008, 5, 1-14.
[16] Andersson, O. Thermal Energy Storage for Sustainable Energy Consumption; Paksoy, H. O. Ed., Springer, Dordrecht, The Netherlands, 2007, 155-176.
[17] Nagano, K; Mochida, T; Ochifuji, K. Appl. *Thermal Eng.*, 2002, 22, 1299-1311.
[18] Fan, R; Jiang, Y; Yao, Y; Shiming, D; Ma, Z. *Energy*, 2007, 32, 2199-2209.
[19] Center for Petroleum and Geosystems Engineering; UTCHEM-9.0 A Three-Dimensional Chemical Flood Simulator; U. of Texas at Austin: Austin, TX, 2000.
[20] Liu, J; Delshad, M; Pope, GA; Sepehrnoori, K. *J. Transport Porous Media*, 1994, 16, 1-29.
[21] Nassar, Y; ElNoaman, A; Abutaima, A; Yousif, S; Salem, A. *Renewable Energy*, 2006. 31, 593-598.

In: Aquifers: Types, Impacts and Conservation ISBN: 978-1-61942-091-5
Editors: Z. Ouakili, et.al, pp. 169-259

Chapter 9

FLUID FLOW AND CONTAMINANT PROPAGATION IN FRACTURED ROCK AQUIFERS

Claudia Cherubini
Department of Civil and Environmental Engineering, Politecnico di Bari, Bari, Italy.

ABSTRACT

It is known that there is a certain unpredictability associated with fractured and karstified rock formations that limits the comprehension of flow and transport phenomena that take place within them. Yet the wide spreading of these rocks makes it extremely important to improve their understanding in order to protect them from natural and anthropic hazards. In fractured rock aquifers the pronounced heterogeneity of the medium due to the existence of discontinuities may create hydrodynamic conditions of difficult interpretation and therefore imply uncertainty in modeling and in planning remediation interventions.

In such context the Darcy law can be adapted to flow throughout fractures by means of the well known cubic law, valid under the assumption of ideal fractures, represented by smooth and parallel plates; however its application to real fracture systems characterized by rough and variable apertures has many times proved to be oversimplified and unrealistic. The parallel plate model is no more valid at higher depth where, due to compressive stresses, the walls of the fractures are pressed together forming asperities that lead to reduction in fracture aperture and cause flow channeling. In addiction, the partial closure of fractures lead to other relevant phenomena in the vadose zone: unsaturated portions within the fractures will act as barriers to liquid flow along them in that the water will tend to flow from one matrix block to another across the partially saturated fractures. In the vadose zone, the hydrodynamic structure of a fractured rock aquifer may be totally reversed: high flux zones at near saturation become low flux zones in unsaturated conditions.

At full saturation however, in fractured aquifers joints act as major conduits for water, dissolved matter and contaminants; in highly heterogeneous fracture networks, open fractures as well as bedding planes, karstic conduits or faults may give even place to preferential flow paths for ground water, contaminants in solution, and free product.

The relatively large size of fracture openings permit fluid to reach high velocities and to move quickly away from a leak site, contaminating large areas in a short time. Furthermore, as contamination moves through a fractured rock aquifer, it tends to diffuse from the flowing fracture water into the rock's essentially stagnant pore water. The rock matrix itself is often able to store ground water, contaminants in solution, or free product.

These processes tend both to retard the plume's advance through a fractured rock mass and to substantially increase the difficulty of purging contamination from the aquifer: the cleanup of fractured rock aquifers in some cases requires many decades, even centuries. The parameters which most strongly govern the degree to which matrix diffusion prolongs the aquifer restoration process have been detected as rock's matrix porosity, fracture spacing, matrix diffusivity, the chemical identity of the contaminant(s), and the length of time the aquifer has been contaminated. It is straightforward that an ad-hoc cleanup technique has to act on the first three parameters that have been identified as crucial within the remediation process itself. Moreover, in fractured rock aquifers numerical models used to simulate remediation processes make use of averaged values for these parameters, as it is extremely difficult to determine them accurately. Making predictions of actual cleanup times is therefore most of the times of scarce significance.

As planning effective remediation interventions requires spatial and temporal predictions on contaminant propagation, it proves to be necessary to implement specific interpretations of the classical methodologies of describing fluid flow and solute transport in fractured rock aquifers.

The present chapter provides an analysis of the peculiar aspects that concern the dynamics of groundwater circulation and contaminant propagation in fractured rock aquifers for the application of the ad-hoc cleanup technique.

1. GROUNDWATER PROPAGATION IN FRACTURED ROCK AQUIFERS

1.1. Fracture Morphology and Dimension

Fractured rock aquifers such as limestone, siltstone, shale, basalt and granite are constituted by a rock matrix inside which the presence of discontinuities is detectable; the term *fracture* is usually referred to any type of discrete discontinuity in rock, of crack, crevice, rift or fissure along which there is hardly any visible displacement (Singhal & Gupta, 1999) parallel or perpendicular to the discontinuity surface; otherwise it is classified as a *fault*.

In geological terms for fracture it is meant any discontinuity that has been originated in consequence of a fragile deformation process of the earth crust. In geomechanical terms, fractures are planes along which stress has caused partial loss of cohesion in the rock. Therefore in rock mechanics, fractures are classified in three macrocategories in terms of relative displacement of fracture walls that has given rise to their own formation:

- Opening fractures (Mode I, Atkinson, 1987), in which the displacement of fracture walls is due to a mechanical stress that is normal to the fracture plane (figure 1.1a).
- Sliding fractures (Mode II, Atkinson, 1987), in which the two fracture walls are displaced in a direction parallel to the plane of the fracture and perpendicular to the fracture front (figure 1.1b)
- Tearing fractures (Mode III, Atkinson, 1987), in which the two fracture walls are displaced in a direction parallel to the plane of the fracture and parallel to the fracture front (figure 1.1c)

Both Modes II and III are also known as *shearing* modes, as fractures are subjected to a shear stress parallel to the fracture plane. Mode II is also termed *In plane shear* fracture and mode III *Antiplane shear* fracture.

The most commonly detectable types of discontinuities are the Mode I fractures; they are also termed joints and can provide channels (or an integrated network of conduits interconnected to each other) for the movement of fluid mass and solute within the rock formation.

Fractures can be natural (generated by tectonic stresses, landslides or movement of glaciers, erosive, shrinkage and weathering processes), or they can also be induced by human activities such as drilling, coring, excavating and blasting (Sara, 2003).

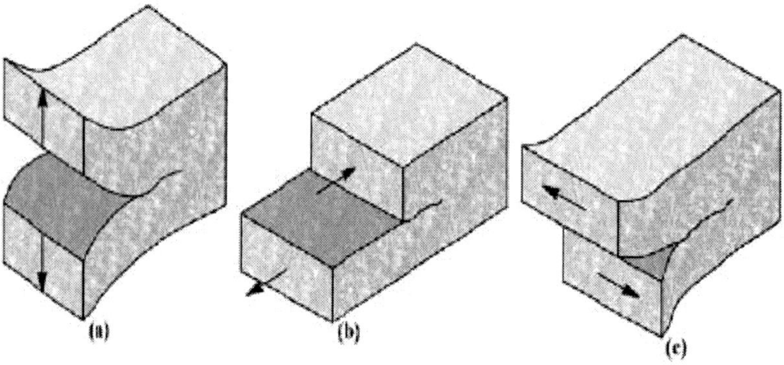

Figure 1.1. type of displacements across fracture: a) Mode I: Perpendicular to fracture, perpendicular to fracture front; b) Mode II: Parallel to fracture, perpendicular to fracture front c) Mode III: Parallel to fracture, parallel to fracture front.

On the basis of their dimensions fractures can be distinguished and therefore can be attributable to one of the following scales:

- Microscopic scale. In crystalline and sedimentary rocks the presence of isolated or continuous microfractures is detectable (mainly joints or to a lesser extent II Mode fractures), prevalently of disk-shaped form, that reach from 100 μm up to maximum 1000 μm length and some μm width;
- Mesoscopic scale: at a higher scale, fractures (mainly joints or to a lesser extent III Mode fractures) with dimensions between 50 cm and 1000 m length and 10 μm and 10 mm width are detectable;
- Megascopic scale. At a even higher scale dominates the presence of shearing fractures (faults), generated by a large scale tectonic displacement parallel to the plane of the discontinuity. Faults can be distinguished by high or low hydraulic conductivity values, depending on local mechanical stresses and on the fault breccia. In relation to their orientation respect to the downflow direction, they can act as barriers or conduits to groundwater flow.

1.2. Characterizing Fractures

Fracture (or fracture systems) are characterized by means of the determination of geometric factors such as aperture, orientation, density and intensity, persistence, connectivity, spacing, trace length, surface roughness and by the presence of wall coatings and infillings along the discontinuity surface.

A variety of hydrogeological investigative techniques can be adopted to characterize fracture systems. Indirect means are constituted by seismic methods, plus other geophysical and geological surveying techniques. Fracture properties can be measured also directly in mine openings, surface outcrop exposures, and on core samples. However, results need to be interpreted with some caution, because they may be affected by stress release (Cook, 2003). Methods for characterizing fractures are described in LaPointe and Hudson (1985) and Singhal and Gupta (1999).

1.2.1. Aperture

The aperture of a fracture is the spacing between its adjacent rock walls and represents one of the main factors controlling fluid flow (Indraratna & Ranjith, 2001). Its magnitude can be affected by external stress, fluid pressure and geometrical properties. Figure 1.2 shows different types of irregularities in joint surfaces, such as variable aperture rough joint with (b) or without (a) closed contacts at discrete points, and with filling material (c).

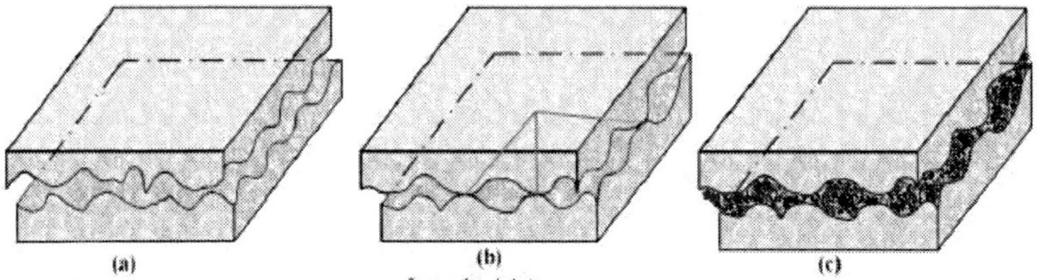

Figure 1.2. Different types of irregularities in joint surfaces (modif from Indraratna & Ranjith, 2001).

Joint aperture can be described in terms of its mechanical or its local aperture. Mechanical aperture (b_M) refers to the separation between the mean levels of the fracture surfaces, whereas local aperture refers to the local spacing between the surfaces and is also known as hydraulic aperture (b_H)(Pawley, 2002) (figure 1.3).

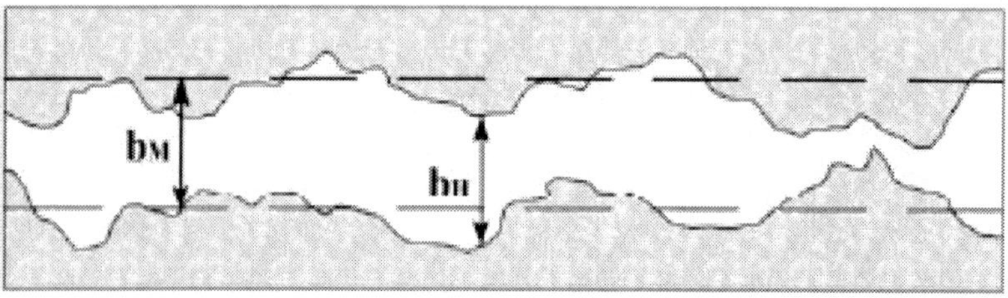

Figure 1.3. Mechanical and hydraulic aperture (Pawley, 2002).

Fracture aperture can be measured along fracture traces on outcrops, cores or well logs. A classification of fractures according to their aperture size is reported in table 1 (Barton, 1973). They can vary from very tight to wide.

Table 1. Aperture classification by size (Barton, 1973)

Size (mm)	Classification
<0.1	Very tight
0.1-0.25	Tight
0.25-0.50	Partly open
0.50-2.50	Open
2.50-10.0	Moderately wide
>10.0	Wide

Several field and laboratory investigations suggest that the distribution of fracture apertures is well described by a lognormal distribution (Pawley, 2002). In contrast, Barton and Hsieh (1989) have suggested a power law distribution for fracture apertures at the site of Yucca Mountain.

However, some caution needs to be applied when interpreting measurements of fracture apertures from surface excavations because they may have been affected by the release of overburden pressures (Cook, 2003). According with Walsh (1981) an increase in compressive stresses leads to a decrease in fracture aperture; this results in a smaller cross sectional area and a higher number of contact points and therefore a higher resistance to flow, because of the longer and tortuous fluid path. Walsh (1981) also showed how an increase in fluid pressure leads to an increase in fracture aperture; as a consequence the area of contact between the surfaces decreases resulting in increased fracture permeability.

Figure 1.4 shows experimental data coming from in situ permeability tests carried out in Colorado by Snow (1968).

At shallow depths (up to 10 m) fracture aperture can vary between 50 and 350 μm; however, at greater depths (more than 30 m), due to compressive stresses, the range of values decreases to 40 - 100 μm.

1.2.2. Orientation

Fractures are commonly presented as lines or arcs in 2D and rectangular planes or circular surfaces in 3D. Fracture orientation can be measured directly from outcrop or underground excavations if the fracture surfaces are exposed (National Research Council, 1996). Otherwise, fracture orientation data from boreholes find widespread use in investigations of the subsurface, though a considerable bias is introduced by the orientation of the borehole or the surface from which the fracture data are collected (Cook, 2003). Fractures parallel to the borehole are not often intersected by the borehole and therefore not sampled. Methods for correction of this bias are discussed by LaPointe and Hudson (1985).

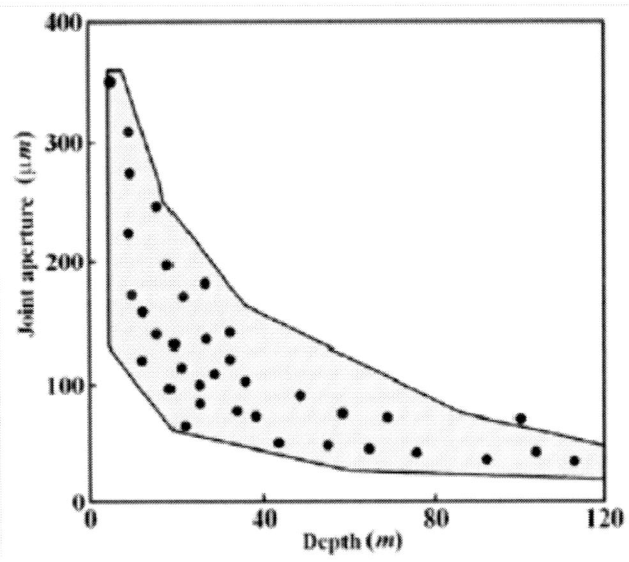

Figure 1.4. fracture apertures at different depths (Snow, 1968).

The orientation in space of a fracture is determined by two angles (figure 1.5):

- dip is the inclination of fracture plane from horizontal (between 0^0 and 90^0)
- azimuth is the angle formed by the fracture line of maximum slope projected on the horizontal and the North (between 0^0 and 360^0)

The orientation is expressed in terms of a pair of numbers, such as 15°/N 250°, implying a plane dipping at 10° in the direction 250° measured clockwise from the north.

In case of 2D fractures, a fracture line is defined by coordinates of the centre (x,y), orientation (ϑ), length (L) or radius (R) (figure 1.6). Fracture orientation is measured from the y-axis (north direction), positively clockwise. The orientation value is equivalent to azimuth angle in 3D case. (Tran, 2004).

1.2.3. Fracture Density and Intensity

Two parameters that play an important role in fluid flow propagation are fracture density and fracture intensity. Accordingly with the definition of Tran (2004) for fracture intensity it is meant the number of fracture counts per unit area or unit volume. Anyway, this definition doesn't take into account that high-intensity fracture zones may not mean also high flux zones, as many small fractures can be not interconnected and therefore may not participate to flow.

On the other side, the parameter fracture density takes into account not only the number of fractures but also their dimensions, in that it is defined in 2D as total fracture length per unit area (A) and in 3D total fracture area per unit volume (V).

$$\text{in 2D: } d = \frac{2 \cdot \sum_i R_i}{A} \tag{1}$$

$$\text{in 3D: } d = \frac{\pi \cdot \sum_i R_i^2}{V} \tag{2}$$

According with Dershowitz and Herda (1992) this definition has proved to be the most useful density measurement in order to characterize discrete fractures.

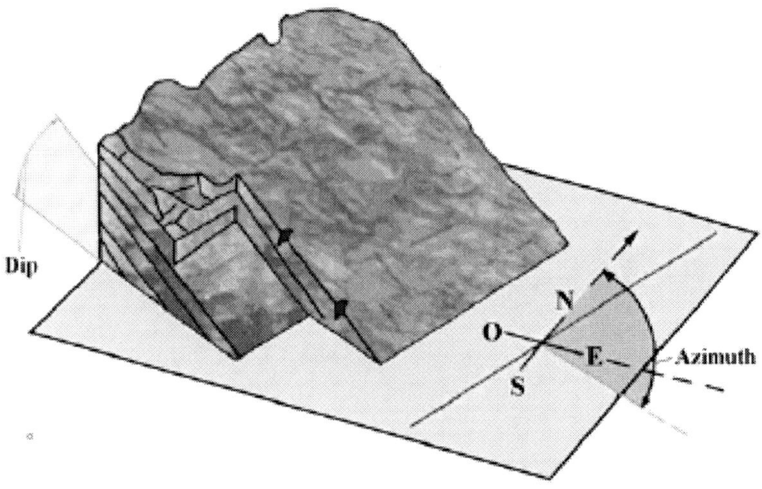

Figure 1.5. 3D fracture orientation.

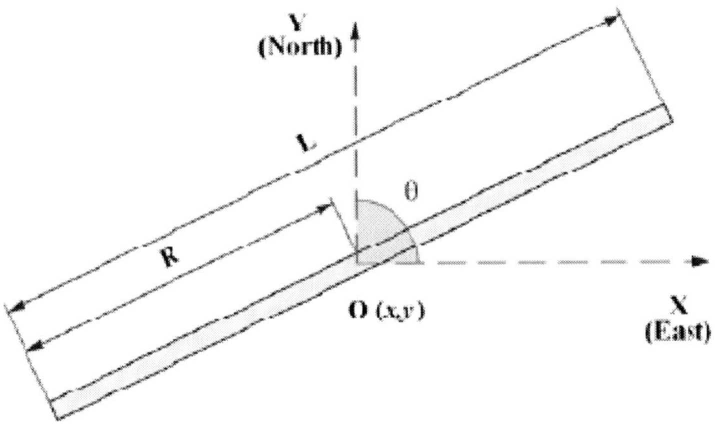

Figure 1.6. 2D fracture orientation (modified from Tran, 2004).

1.2.4. Fracture Spacing

Fracture spacing has been termed by Parsons (1966) as the average perpendicular distance between two adjacent regularly spaced fractures of the same set.

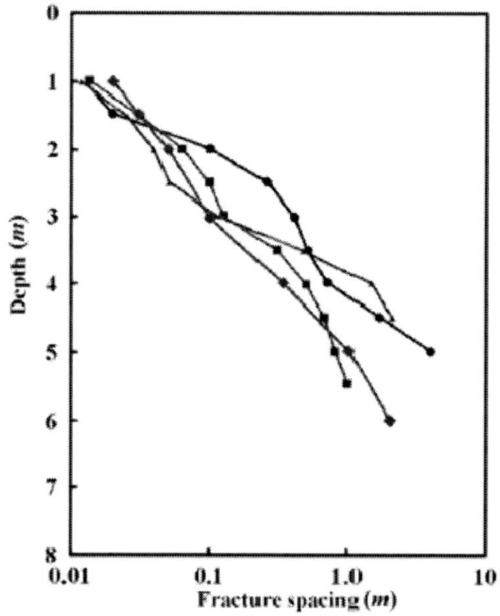

Figure 1.7. Variation of fracture spacing with depth at four sites in glacial till in southern Ontario. (From Ruland et al., 1991).

Obviously this term can be applied to different fracture sets of various orientations within the aquifer. Fracture spacing is influenced by rock composition, texture, structural position, and bed thickness. Variation in fracture spacing can have a dramatic effect on both fracture porosity and permeability (Nelson, 1985), as It will be shown in the next paragraphs. Within units of the same lithology, an increase in fracture spacing with depth (figure 1.7) is frequently observed (Cook, 2003).

1.2.5. Persistence and Trace Length

The persistence of a joint represents the areal extension or the dimension of a discontinuity within a plane and is therefore defined, with reference to the plane on which the joint lies, as the fraction of area belonging to the discontinuity itself (figure 1.8):

$$p = \lim_{A \to \infty} \frac{\sum a_{ti}}{A_t} \cong \frac{\text{area of the joint}}{\text{total area}} \tag{3}$$

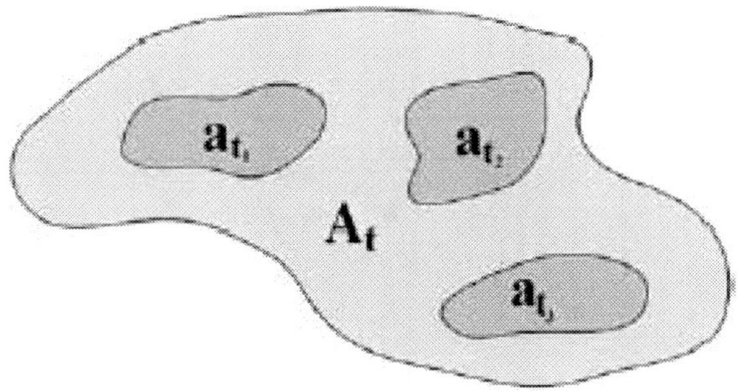

Figure 1.8. Joint persistence expressed in terms of areas.

In the same way, in terms of length, the persistence can be defined as the limits of the ratio between the sum of the discontinuous lengths of a given trace of the plane itself and the total length:

$$p = \lim_{L \to \infty} \frac{\sum l_{si}}{L_s} \tag{4}$$

Where L_S is the length of the segment S of a straight line of a trace of the plane and l_{si} is the length of the i joint in S.

Trace length is the observed length of a fracture trace and can be observed in outcrops or in excavations. The observed trace length may be only an apparent value of the true trace length due to various types of bias creeping into the data during measurements in exposures. Fracture length can be as small as wellbore scales (less than a metre) or as big as reservoir scales (more than a few kilometers) (Tran, 2004). In a literature review performed by Baecher et al. (1977) and in a successive study developed by Baecher and Lanney (1978), joint trace length is reported to vary lognormally.

1.2.6. Fracture Connectivity

Parameters such as orientation, density, length and spacing are important determinants of the degree to which fractures are interconnected; the degree of interconnectedness of fracture sets is a critical feature contributing to the hydraulic conductivity of a fractured rock system (Witherspoon et al., 1987) and therefore to the ability of fractures to act as conduits for groundwater flow. The importance of joint connectivity in a fractured rock aquifer is represented in figure 1.9. In figure 1.9a, although there is the presence of numerous clusters of fractures, none of them are interconnected and consequently prevent groundwater flow through the system. Contrarily, in figure 1.9b the interconnected clusters of fractures provide effective pathways for groundwater flow through the system.

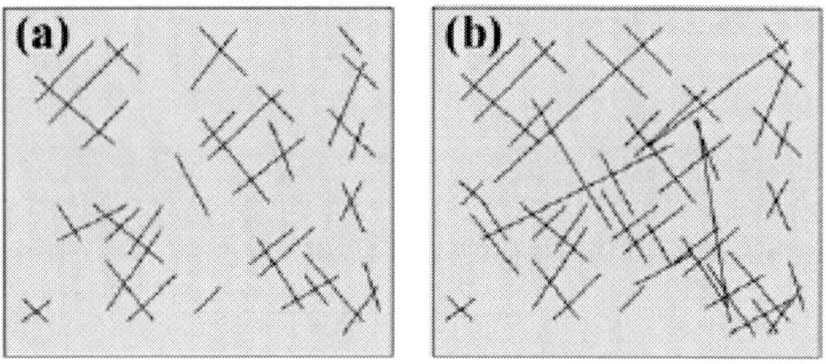

Figure 1.9. Fracture connectivity, (a) not interconnected clusters of fractures; (b) interconnected clusters (Domenico & Schwartz, 1998).

Fracture connectivity increases with increasing fracture length and density, as the chance of fracture intersection increases (Cook, 2003) (figure 1.10). Singhal and Gupta (1999) expressed interconnectivity in terms of the ratio of fracture trace length to average fracture spacing.

According with their degree of connectivity, Barton et al. (1987) classified fractures into three categories: Crossing, abutting and blind. In figure 1.10a both sets of fractures are characterized by fracture length large relative to fracture spacing: all fracture terminations are *crossing*; On the other hand, in figure 1.10b one set of fractures has fracture length greater than fracture spacing, while the other set has fracture spacing greater than fracture length. In this case all fracture terminations are *abutting*; figure 1.10c shows both sets of fractures that have fracture length smaller than fracture spacing: in this circumstance all fracture terminations are *blind* (Cook, 2003).

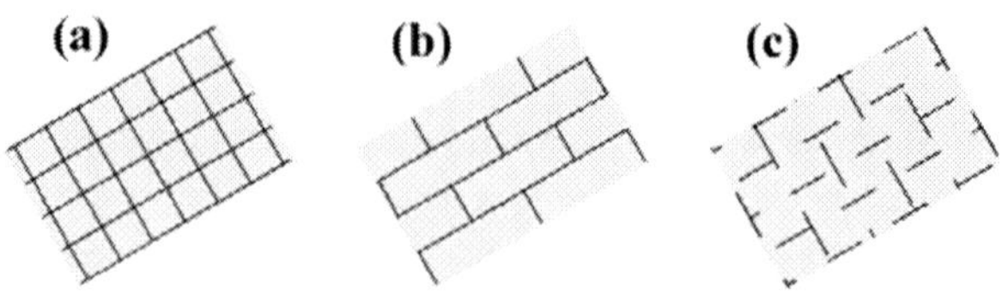

Figure 1.10. Fracture different degree of connectivity (modif from Singhal & Gupta, 1999).

1.2.7. Surface Roughness

Fracture walls are not flat parallel smooth surfaces, but contain irregularities called asperities (Singhal & Gupta, 1999). The effect of roughness varies with the change of joint aperture, thickness of infill material and the relative displacement of joint walls (Idraratna & Ranjith, 2001). A quantification of the degree of roughness based on unevenness and waviness of joints has been given by ISRM commission (1978) (figure 1.11a).

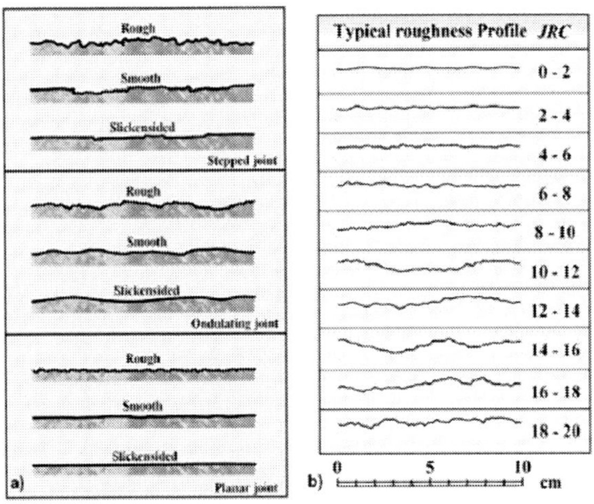

Figure 1.11. a) various degrees of roughness (ISRM, 1978); b)Typical fracture roughness profiles (Barton & Choubey, 1977).

Surface roughness is measured by comparing the profile of an exposed, unweathered fracture surface with a standard set of profiles. Each of the ten 100 mm long profiles is attributed a range of *Joint Roughness Coefficient* (*JRC*) from 0 to 20, representing a sequential increase in roughness (Barton & Choubey, 1977; figure 1.11b).

Barton and Bandis (1980) proposed a relationship that accounted for the scale effect between the JRC standard profiles (JRC_S) for a 100 mm sample length (L_S) and the equivalent roughness of a profile of length L:

$$JRC = JRC_s \left(\frac{L}{L_s} \right)^{-0.02 JRC_s} \tag{5}$$

This correction is necessary in case of samples characterized by a length higher than 100 mm, in order to take large- scale roughness into account (Price, 2005). As illustrated in figure 1.12, surface roughness leads to discrepancy between b_H and b_M. for low JRC values, $b_H \approx b_M$ (100 μm), while it decreases steeply with raising roughness. Barton et al. (1985) proposed an empirical relationship between b_H, b_M and JRC, in which aperture measurements are in micron:

$$b_H = \frac{b_M^2}{JRC^{2.5}} \tag{6}$$

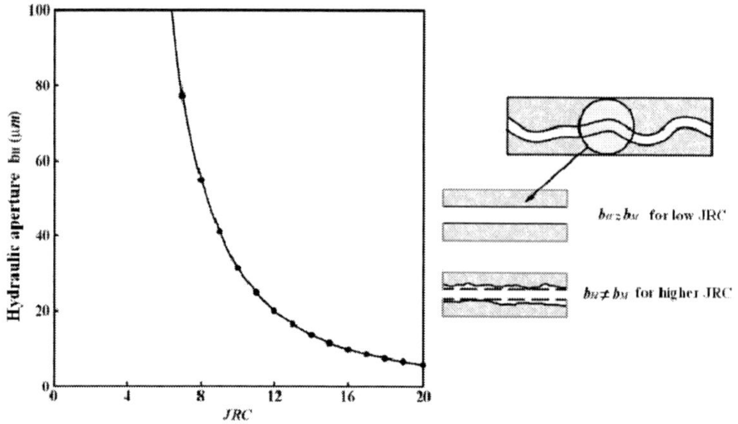

Figure 1.12. Relationship between hydraulic aperture (***b****H*) and JRC for a hypothetical fracture with mechanical aperture b_M of 100 μm (modif from Price, 2005).

1.2.8. Wall Coatings and Infillings

This notation is referred to clay, terra rossa, fault gauge, breccia, chert, calcite and any other solid material occurring along the discontinuity surface that can partly or completely fill the discontinuity itself. Filling material can be homogeneous or heterogeneous and can have variable permeability and porosity, depending on mineralogy, grain size, etc (Singhal & Gupta, 1999). Fracture and joints filled with secondary minerals are often referred to as *veins* (Sara, 2003).

1.2.9. Hydraulic Conductivity and Porosity of Fractures and Matrix

Fractured rock aquifers are constituted by a network of discontinuities within a rock matrix, therefore their characterization requires information on the nature of both fractures and matrix. The latter can be characterized by its pore distribution, that can be expressed in terms of porosity and hydraulic conductivity.

Groundwater flow takes place throughout a combination of intergranular pores and interconnected fractures. Though both fractures and matrix contribute to groundwater flow, due to the low hydraulic conductivity of the latter in many fractured rocks, the predominant mechanism is constituted by fracture flow; therefore, in several practical applications, in cases in which fracture hydraulic conductivity is about orders of magnitude higher than matrix hydraulic conductivity, the effect of the latter can be neglected.

Matrix hydraulic conductivity is directly linked to the porosity of the material itself that depends on factors such as lithology, geologic history and the in situ stress conditions. Fracture hydraulic conductivity is function of their distribution and geometry (aperture, spacing, connectivity etc etc).

The equivalent hydraulic conductivity of a rock mass (K_r) with one parallel set of fractures can be expressed as (Singhal, 2008):

$$K_r = \frac{b}{B} K_f + K_m = \frac{b^3}{B} \frac{\rho g}{12\mu} + K_m \tag{7}$$

Where b is fracture aperture and B fracture spacing, K_f and K_m represents fracture and matrix hydraulic conductivities. Usually K_m is very low in comparison with k_f except when rock is porous and/or fractures are filled with impervious material (Singhal, 2008).

Porosity *n* is the ratio between the volume of voids present inside the rock formation and the total volume. Fractured rock masses are characterized by a two-type porosity: the primary -also called intergranular- porosity, or matrix porosity, is linked to the intrinsic characteristics of the rock due to its formation. The secondary porosity, also termed fracture porosity, develops successively due to different geological processes.

Fracture porosity (n_f) is defined as the volume of aquifer occupied by open fractures in relation to the total volume; normally it has low values (always less than 2%, generally less than 0.5%), but can also assume a relevant role in aquifers characterized by high extent both spatially and vertically. Matrix porosity (n_m) is always such that $n_m >> n_f$.

Laboratory measurements carried out on a variety of fractured rocks show that fracture porosity generally varies between 0.001% and 0.01%, whereas matrix porosity between 0.1% and 8% (Lee & Farmer, 1993). Nelson (1985) has proposed the following expression for the calculation of fracture porosity:

$$n_f = \frac{b}{b+B} \times 100 \tag{8}$$

In figure 1.13 is showed the variation of fracture porosity and hydraulic conductivity as a function of fracture aperture and spacing. As it can be pointed out from the graphs, both porosity and hydraulic conductivity values raise with an increase in fracture aperture and decrease with an increase in fracture spacing, as Eqq [7] and [8] indicate.

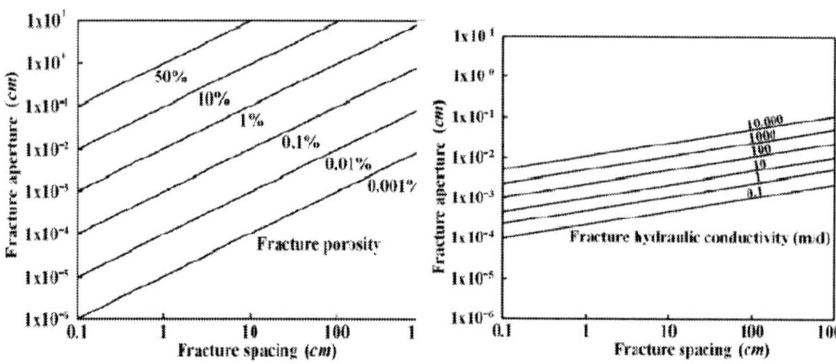

Figure 1.13. a) fracture porosity and b) hydraulic conductivity as a function of fracture aperture and spacing (modified from Nelson, 1985).

Fracture and matrix porosity have a different influence on hydraulic conductivity. As far as fractures, a minimum increment of n_f correspond to a remarkable increase of k_f (Nelson, 1985). The straight line belonging to non fractured sandstone may be representative of the matrix, for which there is a more gradual increase of hydraulic conductivity with porosity.

1.2.9.1. Variation of Hydraulic Conductivity with Depth

Several in situ measurements have detected a decrease of hydraulic conductivity with depth in fractured rock masses, attributable to a reduction of fracture aperture due to an increase in loading. In this context, different empiric laws have been developed that generally provide corrective factors for shallow hydraulic conductivity that account for depth. In a recent study (Clerici & Sfratato, 2008) it has been demonstrated that the empirical relations based only on depth from the ground surface return values that fit better the ones obtained by in situ tests

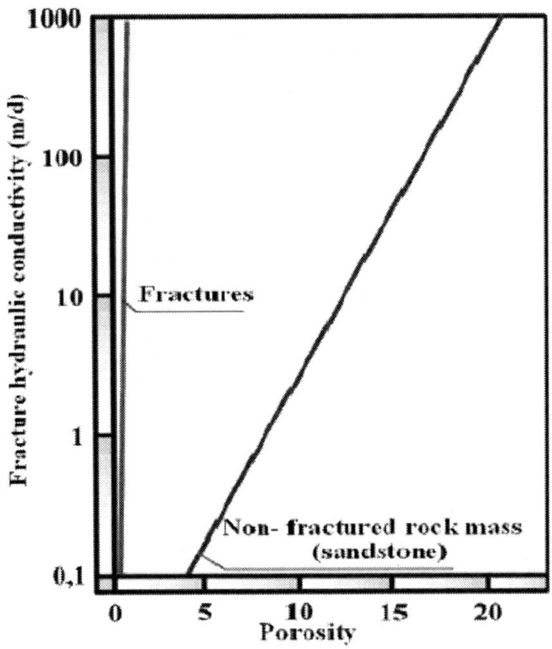

Figure 1.14. Hydraulic conductivity as a function of porosity for fractures and non fractured rock mass (sandstone) (modif from Nelson, 1985).

The dependence of permeability on depth can be expressed by the relation of Black (1987):

$$k = aZ^{-c} \tag{9}$$

Dove:

a, c= constants
Z= depth [m]
k= permeability [m^2]

The relation of Snow (1968), proposed for the calculation of permeability in crystalline rock masses depends uniquely on depth:

$$k = 10^{-(1,6 \log Z + 4)} \tag{10}$$

Carlsson & Olsson (1977) propose two relations for the calculation of the variation of hydraulic conductivity with depth, also dependent only on Z:

$$K = 10^{-(0,002 Z + 5,6)} \tag{11}$$

$$K = 10^{-(2,5 \log Z + 2,5)} \tag{12}$$

Where:

K= hydraulic conductivity [m/s]
Analogously, Burges relation (In Lee & Farmer, 1993):

$$\log K = 5,57 + 0,352 \log Z - 0,978(\log Z)^2 + 0,167(\log Z)^3 \tag{13}$$

Higher permeabilities at shallow depths (<50 m) can be attributed also to greater influence of surficial phenomena such as weathering, stress relief caused by erosion and unloading phenomena. Nevertheless, a decrease in permeability with depth due to increased stress may not be systematic (Singhal, 2008), in that the permeability itself can vary several orders of magnitude at the same depth (figure 1.15).

Figure 1.15 represents the permeability measured in short interval well tests in fractured granitic rock at Gideå, Sweden (Rutqvist & Stephansson, 2003). A general decrease in permeability with depth is observed, but the depth dependency is obscured by a large permeability spatial variation (up to 6 orders of magnitude) (Rutqvist & Stephansson, 2003), due to other intervening factors such as mineral precipitation/dissolution and shear dislocation. Although generally in fractured rock aquifers a decrease in permeability due to increase in compressive stresses is observed, there is not much justification of such a universal rule (Singhal, 2008).

The lowest values on the left hand side of figure 1.15 represent the permeability of the rock matrix: no fractures are intersecting or, if so, they are completely cemented with minerals or not interconnected and therefore not conducting. On the other side, the highest values on the right hand side of figure 1.15 are representative of highly transmissive, interconnected and therefore conducting fractures.

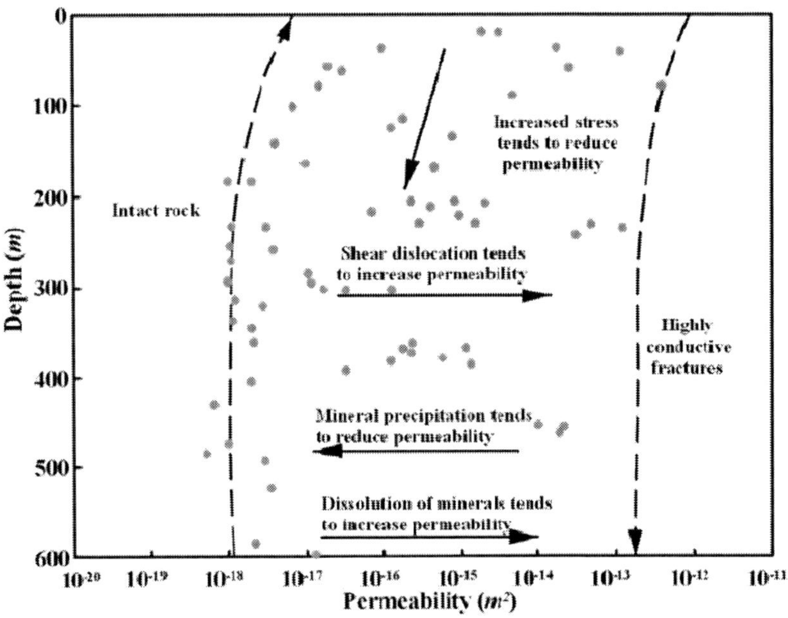

Figure 1.15. Variation of permeability with depth at Gideå, Sweden (Rutqvist & Stephansson, 2003).

1.3. Representative Elementary Volume (REV)

1.3.1. Definition of REV

The Representative Elementary Volume (REV) is defined as the smallest sample volume that contains the intrinsic characteristics of the rock mass. This volume varies with the type of rock and fractures, and in some circumstances is even indefinable.

The concept of REV is necessary in order to define the distribution of parameters that characterize the aquifer, such as for example hydraulic conductivity. In homogeneous non fractured rocks hydraulic conductivity tends to assume a constant value beyond a certain sample volume, called Representative Elementary Volume (figure 1.16a). In a fractured rock mass, hydraulic conductivity may not assume a constant value with the increase of the examined volume even if, in some cases, the variations can be considered negligible (figure 1.16b).

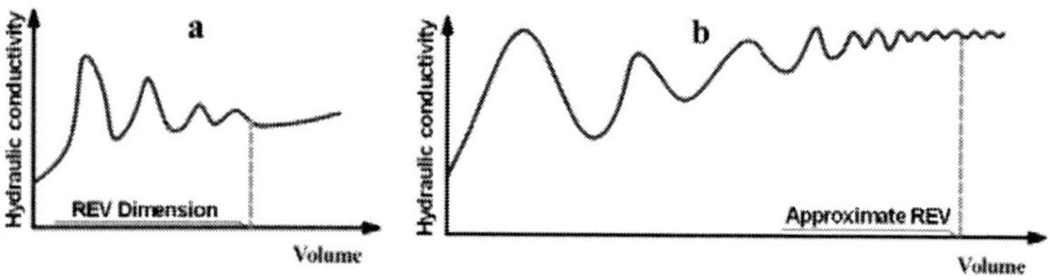

Figure 1.16. REV for: a) for porous media; b) for fractured media (modif from Singhal & Gupta, 1999).

The concept of REV is applicable to fractured rocks when particular conditions are met, that make it assimilable to a homogeneous anisotropic medium, that is to say: a constant hydraulic gradient together with a linear flow system.

The Representative Elementary Volume increases with the increase of the sizes of the discontinuities present in the site. The influence that the geometry of the system of fractures exerts on REV is shown in figure 1.17. In rocks that do not show any relevant discontinuities, a small REV can be retained representative of the entire rock mass (figure 1.17a), whereas in fractured rocks the REV dimensions must be large enough to contain sufficient intersections between fractures in such a way as to represent the flow domain (figure 1.17b).

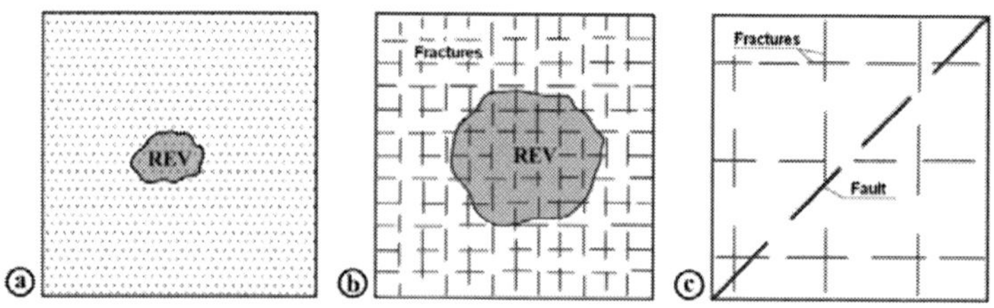

Figure 1.17. REV in different rock masses: (a) homogeneous porous rock, (b) fractured rock in which REV contains sufficient fracture intersections and (c) rock with large scale discontinuities in which REV is too extended or doesn't exist (modified from Singhal & Gupta, 1999).

The REV dimensions have to be sufficiently large in relation to the length of the fractures in such a way as to be able to enclose a representative sample of the fracture population.

Renshaw (1998) argues that the maximum hydraulic conductivity occurs at the scale that is just great enough for a single large cluster of fractures to form that spans the entire network (Renshaw, 1998).

According to Odling et al. (1999) a REV in a fractured rock mass must have dimensions ranging from one to two orders of magnitude higher than the average fracture spacing.

However, in fractured rock aquifers characterized by greatly variable fracture systems, in terms of length and spacing, a definition of REV may be a complicated task. Some researchers argue that hydraulic conductivity continues to increase with the scale of investigation because the probability of intersecting larger fractures increases. According to others, at larger scales, hydraulic conductivity begins to decrease as fracture connectivity is reduced, due to fractures having finite lengths (Cook, 2003).

Anyway, in case of large scale discontinuities, such as faults, the individuation of REV may not be technically possible because it will involve a too large area (Singhal & Gupta, 1999) (figure 1.17c). This means that the REV concept cannot be always applicable to fractured rock masses.

1.3.2. Scale Effect in Fractured Rock Hydraulic Parameters

The term scale effect, or scale dependence, represents the fact that hydraulic properties in fractured rock aquifers depend on the volume of rock though which flow is occurring. Different methods of measurement of hydrodynamic parameters correspond to different scales of investigation and therefore might lead to different results.

Fracture hydraulic conductivity and porosity, having a spacing term in their calculation, are scale dependent and as such present more severe sampling problems than matrix hydraulic conductivity and porosity do (Nelson, 1985).

As far as hydraulic conductivity is concerned, methods for determining it are classified into three basic types: laboratory tests, borehole tests and calibration of regional or basin- scale modeling (Hsieh, 1998).

In a system of evenly spaced, identical fractures, at very small scales the hydraulic conductivity oscillates between that of the matrix, Km, and that of the fractures, Kf (figure 1.16b)

However, in laboratory hydraulic conductivity measurements are usually made from compact and unfractured core samples since it is technically relatively difficult to collect samples within fractured zones; they are therefore more representative of the matrix, showing lower values.

On the other side, in situ permeability values are significantly greater and more representative as in the boreholes a greater thickness of rock with fractures is sampled; however, these values show also a greater variability depending on the aperture and frequency of fractures intercepted by the borehole (Singhal & Gupta, 1999).

When measurements are made at scales much larger than the fracture spacing, then the variability of hydraulic conductivity will be greatly reduced and a value equal to the aquifer hydraulic conductivity will be determined. Therefore it proves to be necessary to use a large enough sampling element to encounter several regularly spaced fractures in order to get accurate measurements (Nelson, 1985).

Different tests carried out on Marville Formation (Wisconsin, USA) by Schulze-Makuch & Cherkauer, (1995) show an increase in hydraulic conductivity with scale (figure 1.18) to a certain point when it levels off and the average hydraulic conductivity is constant with scale. This levelling off is a homogenation effect due to proximity to REV.

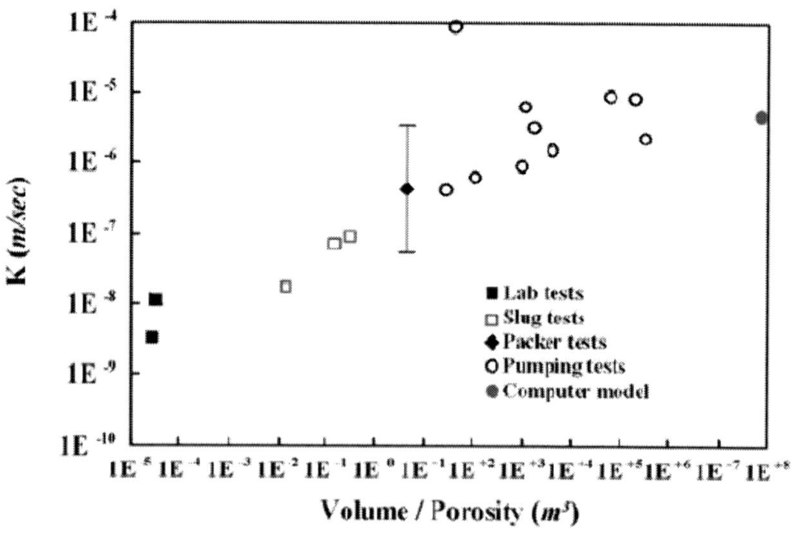

Figure 1.18. Scale dependency of hydraulic conductivity of the vuggy Mayville facies (Schulze-Makuch & Cherkauer, 1995).

Data obtained from different fractured rock aquifers by Clauser (1992) suggest that permeability increases up to a homogenization scale of about 100 m (REV) beyond which it approaches an approximatively constant value (figure 1.19). The apparent small decrease in conductivity measured by slug tests at regional scale may be attributed to a decrease in fracture aperture due to blocking of fractures by particulate material and the precipitates formed due to chemical differences in the quality of groundwater and injected water (Singhal & Gupta, 1999).

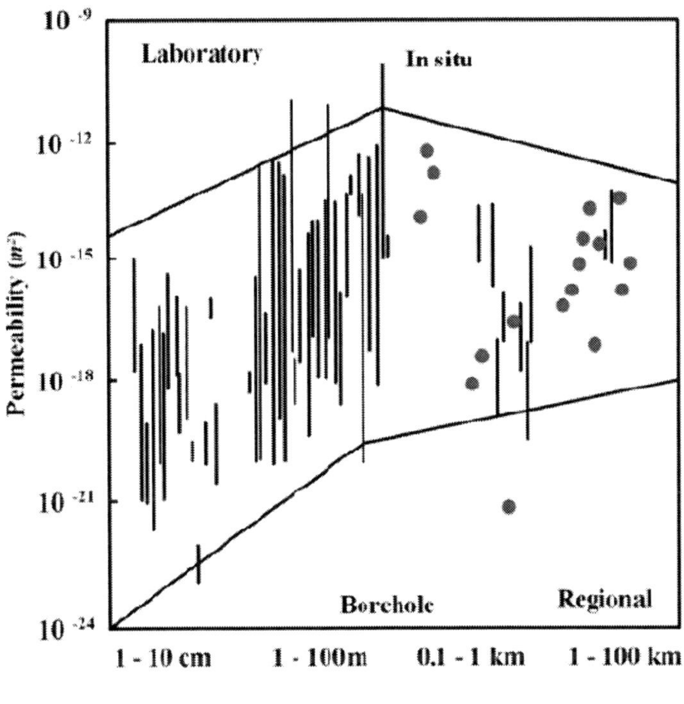

Figure 1.19. range of permeability values of crystalline rocks as a function of scale of measurement (Clauser, 1992).

1.4. Flow in Fractured Rocks

1.4.1. From the Navier - Stokes Equations to the Stokes Equations

Let's consider a two-dimensional fracture whose aperture varies just in x, that is assumed as the flow direction; the direction perpendicular to the fracture plane will be z (Zimmermann & Yeo, 2000) (figure 1.20a). Otherwise if the fracture is three dimensional, the other coordinate inside the fracture plane will be y. Groundwater velocity inside a fracture is higher in the middle part due to the presence of frictional forces along the fracture walls (figure 1.20b). If the hydraulic gradient is constant, the mean groundwater flow velocity through a fracture will increase with the distance from the fracture walls. The mean velocity will be also higher if the fracture walls are flat and smooth rather than irregular and with asperities.

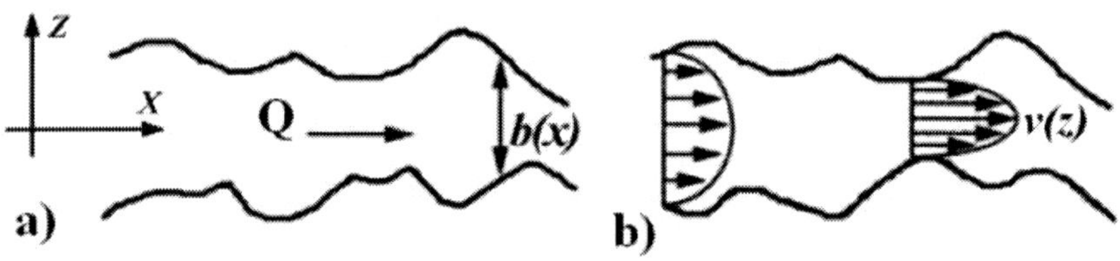

Figure 1.20. a) Two dimensional fracture with aperture variable in x direction; b)parabolic profile of velocity along the fracture (modif from Zimmermann & Yeo, 2000).

The hydrodynamic processes occurring within rock fractures are governed by the Navier-Stokes equation and the continuity equation; the Navier Stokes (N-S) equations are a set of three *non linear* partial differential equations that in their general form can be written as:

$$\rho\left[\frac{\partial v}{\partial t}+(v\cdot\nabla)v\right]+\nabla p-\mu\nabla^2 v+\rho g=0 \tag{14}$$

where V is the velocity vector (v_x, v_y, v_z) and P is the pressure in the moving fluid at each point; ρ and μ are density and dynamic (or absolute) viscosity of the fluid, respectively; and ∇ and ∇^2 are the divergence and the Laplacian, operator respectively; ρg represents gravity, the only appreciable external force per unit volume acting on the flowing fluid where g is the gravitational acceleration vector = 9.81 m/s^2.

It is possible to express the equation in terms of piezometric head (Bear, 1993), defined, for a fluid of constant density, as:

$$H=z+\frac{p}{\rho g} \qquad .(15)$$

In this case:

$\nabla p + \rho g = \nabla[\rho g(H - z)] + \rho g \equiv \rho g \nabla H$ if the density is constant

In the steady state, the Navier Stokes equations can be reduced to:

$$\mu \nabla^2 v - \rho(v \cdot \nabla)v = \rho g \nabla H \tag{16}$$

The first term represents the viscous forces, the second the advective acceleration and the third the pressure gradient. The second term on the left-hand side of the equation makes the equation nonlinear and, consequently, difficult to solve. The Navier-Stokes equations have to be integrated with the equation of mass conservation (continuity equation) that for an uncompressible fluid takes the form:

$$\nabla \cdot u = 0 \tag{17}$$

The previous equations form a system of four equations in the four unknowns (u_x, u_y, u_z, H):

$$\begin{aligned}
&\frac{\partial v_x}{\partial x} + \frac{\partial v_y}{\partial y} + \frac{\partial v_z}{\partial z} = 0 \\
&\rho\left(v_x \frac{\partial v_x}{\partial x} + v_y \frac{\partial v_x}{\partial y} + v_z \frac{\partial v_x}{\partial z}\right) = -\rho g \frac{\partial H}{\partial x} + \mu\left(\frac{\partial^2 v_x}{\partial x^2} + \frac{\partial^2 v_x}{\partial y^2} + \frac{\partial^2 v_x}{\partial z^2}\right) \\
&\rho\left(v_x \frac{\partial v_y}{\partial x} + v_y \frac{\partial v_y}{\partial y} + v_z \frac{\partial v_y}{\partial z}\right) = -\rho g \frac{\partial H}{\partial y} + \mu\left(\frac{\partial^2 v_y}{\partial x^2} + \frac{\partial^2 v_y}{\partial y^2} + \frac{\partial^2 v_y}{\partial z^2}\right) \\
&\rho\left(v_x \frac{\partial v_z}{\partial x} + v_y \frac{\partial v_z}{\partial y} + v_z \frac{\partial v_z}{\partial z}\right) = -\rho g \frac{\partial H}{\partial z} + \mu\left(\frac{\partial^2 v_z}{\partial x^2} + \frac{\partial^2 v_z}{\partial y^2} + \frac{\partial^2 v_z}{\partial z^2}\right)
\end{aligned} \tag{18}$$

Navier-Stokes equations are not solvable in case of real fractures, constituted by rough walls and are difficult to be solved also for ideal geometries constituted by sinusoidal or saw-tooth walls. Therefore it is necessary to make some approximations to reduce these equations to a simpler form. The first level of simplification consists in neglecting the acceleration terms. If the viscous forces dominate the inertial forces (Batchelor, 1967), the

Navier-Stokes equations reduce to the Stokes equation under steady-state conditions:

$$\rho g \nabla H = \mu \nabla^2 u \tag{19}$$

The Stokes equations represent a set of three *linear* partial differential equations.

The reduction of the Navier Stokes equations to the Stokes equations is possible only if the terms of advective acceleration $u \cdot \nabla u$ are smaller than the viscous terms $\nu \nabla^2 u$. Anyway, as it is not easy to estimate the dimension of the single terms without finding the detailed solution to the problem, it is consequently difficult to establish a priori conditions for the Navier Stokes equations to be replaced by the Stokes equations (Zimmermann & Yeo, 2000)

1.4.2. The Parallel Plate Model

The simplest model of fluid flow through a rock fracture is represented by a fracture constituted by flat and smooth walls, separated by an aperture *b* (Sarkar, 2004), that is known as the *parallel plate model* (figure 1.21). It has the sole geometry for which the N-S equations can have an exact solution (Zimmermann & Yeo, 2000).

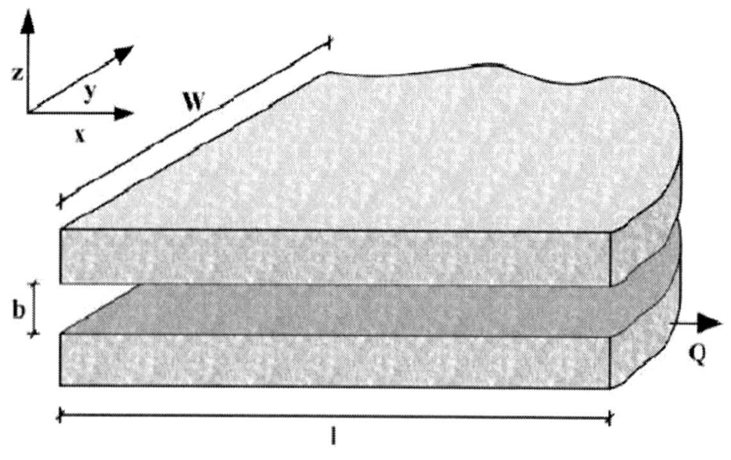

Figure 1.21. Parallel plate model.

In this case the N-S equations, as the velocity vector and the velocity gradient are orthogonal to each other and the non linear advective term vanishes identically.

The velocity vector for flow between smooth, parallel plates of aperture *b* is given by [Batchelor, 1967]:

$$v_x = -\frac{\rho g}{2\mu}\frac{dH}{dx}\left[z^2 - (b/2)^2\right], \quad v_z = 0 \tag{20}$$

This profile can be integrated across the fracture, to find the overall flow rate Q_x (Zimmerman & Bodvarsson, 1996):

$$Q_x = \int_0^w \int_{-b/2}^{b/2} v_x(z)dz\ dy = -w\int_{-b/2}^{b/2} \frac{\rho g}{2\mu}\frac{dH}{dx}\left[z^2 - (b/2)^2\right]dz = \frac{-\rho g w b^3}{12\mu}\frac{dH}{dx} \tag{21}$$

Where *w* represents the width of the fracture, perpendicular to flow direction.

Darcy law for a one dimensional flow trough porous medium can be written:

$$Q = -\frac{KA}{\mu}\left(\frac{dH}{dx}\right) \tag{22}$$

In the current case, the cross sectional area is *wb*. Therefore fracture hydraulic conductivity can be expressed in the following way:

$$K_f = b^2\frac{\rho g}{12\mu} \tag{23}$$

This formulation permits to evidence the different hydraulic properties existing between fractures and matrix. Assuming that the fluid flowing through a fracture is water with a density ρ_f = 1000 kg m^{-3} and viscosity μ_f = 0,001 kg m^{-1} s^{-1}, using the previous equation and substituting the value of gravitational acceleration (g = 9,81 m s^{-2}), the hydraulic conductivity of a single fracture of 250 μm aperture results approximately equal to 0,05 m s^{-1}, value typical of a highly permeable material, as gravel. Instead, the hydraulic conductivity of a compact rock mass where the fracture is allocated, can result almost 10 orders of magnitude lower. It results that the groundwater flow velocities in a single fracture can be several orders of magnitude higher than in the rock matrix. Starting from the [23] it is possible to adapt Darcy law to flow through fractures, making the assumption of ideal fractures with flat and smooth walls and infinite lengths, together with laminar flow and uncompressible fluid. In this way it is possible to obtain (Witherspoon et al., 1980):

$$Q = \underbrace{\frac{b^2 \rho g}{12\mu}}_{K} \cdot \underbrace{b \cdot w}_{A} \cdot i = \frac{b^3 \rho g}{12\mu} \cdot w \cdot i \tag{25}$$

The [25] is called cubic law because the dependence of hydraulic conductivity on fracture aperture is cubic. A doubling of fracture aperture leads to an increment of a factor eight in hydraulic conductivity (figure 1.13b).

In a system of equally distant fractures, flat, parallel in an impermeable matrix, the equivalent hydraulic conductivity (in the direction parallel to the fractures) can be expressed by a relation developed by Snow (1968):

$$K = \frac{b^3}{B} \frac{\rho g}{12\mu} \tag{26}$$

Where B represents fracture spacing. For example, in a fractured medium characterized by fracture spacing $B = 1$ m and aperture $b = 250$ μm, will have a hydraulic conductivity approximatively equal to 10^{-5} m s^{-1}, similar to a coarse sand and will have also the same hydraulic conductivity of a fractured medium with fracture spacing equal to 10 cm and aperture 115 μm (figure 1.22).

In the same way, an equivalent hydraulic conductivity for a set of planar fractures can be calculated with another equation developed by Snow (1968):

$$K = Nb^3 \frac{\rho g}{12\mu} = n_f b^2 \frac{\rho g}{12\mu} \tag{27}$$

where n_f represents fracture porosity and is equal to the product of N and b, N is the number of joints for unit length along the rock wall (L^{-1}). In this case the equivalent hydraulic conductivity is linearly proportional to the product of fracture porosity and the square of fracture aperture.

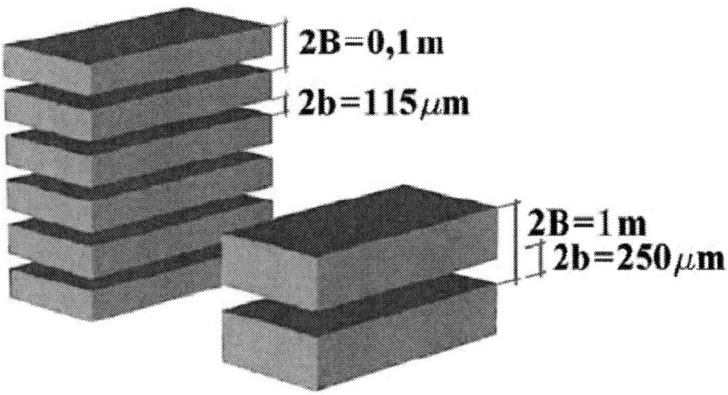

Figure 1.22. Two different rock masses constituted by a system of plane, parallel and uniform fractures having the same hydraulic conductivity (10^{-5} m s^{-1}). a) fractures with aperture of 250 μm and spacing 1 m, b) fractures with aperture of 115 μm and spacing of 0.1 m (modified from Cook, 2003).

In porous media the discharge $q = K \cdot i$ is linked to velocity by the known expression q=V·n, where n represents the porosity of the medium. In a purely fractured medium the discharge is linked to velocity though fractures V_w by means of the following expression:

$$q = V_w n_f = V_w \frac{b}{B} = V_w bN \tag{28}$$

Where, starting from the [8] and neglecting b respect to B it follows that n_f = b/B. While velocities in porous media are usually quite low (typically from 1 to 100 m/y)velocities in fractures can be even very high. They are often expressed in m/d instead of m/y. Assuming a hydraulic conductivity of 10^{-5} m s^{-1} and a hydraulic gradient of 10^{-2}, the Darcy law returns a discharge value equal to 10^{-7} m^3/s, that is to say equal to 3.15 m^3/y. if fracture spacing is 1 m and fracture aperture 250 μm, therefore fracture velocity will be equal to 12600 m/y, that is to say 34,5 m/d. Figure 1.23 shows velocities through fractures of different apertures as a function of hydraulic gradient. Fracture velocity is proportional to the square of aperture, while the discharge is proportional to the cube of aperture (Cook, 2003).

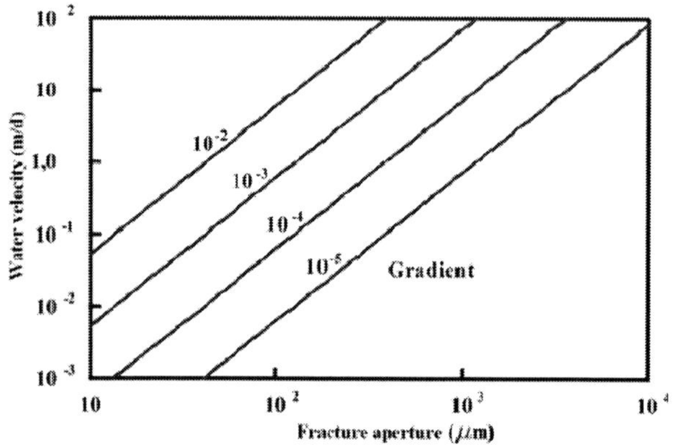

Figure 1.23. Velocity through plane fractures as a function of fracture aperture and hydraulic gradient (modified from Cook, 2003).

The Hagen Poiseuille equation has been also used to estimate average fluid velocity through an isolated cylindrical flow channel under laminar flow conditions (Inyang& Bruell, 2000).

$$V = -\frac{\rho g r^2}{8\mu} \cdot i \tag{29}$$

Where V is the velocity of flow (LT^{-1}), r is the radius of the flow channel (L). Chen and Wagenet (1992) suggest that this equation can be used only in cases in which the Reynolds number (R_e) ≤ 3.

Physical experiments conducted by Bouma et al. (1979) indicate that for tube diameters >100μm, the Hagen Poiseuille equation is inadequate. Thus for transitional and turbulent flow regimes (corresponding to progressively higher Reynolds numbers) the Chezy-Manning and Manning relationships can be used.

1.4.3. Effect of Roughness and Turbulent Flow in Fractures

The cubic law has been derived under the hypothesis of a fracture constituted by parallel plates. Real rock fractures instead, are characterized by rough walls and variable apertures. Moreover, even contact points between two opposite fracture walls are likely to occur. Being transmissivity proportional to b^3, flow in a fracture characterized by variable aperture in saturated conditions will tend to follow pathways of least resistance, that is to say characterized by maximum aperture, differently from what happens in the simplified parallel plate model. In such cases, in order to predict transmissivity of a real fracture, the use of the cubic law is allowed if the aperture b is replaced by an *equivalent aperture*. Therefore, for real fractures, transmissivity is quantified in terms of "*hydraulic aperture*" b_H, that is defined as the aperture that, if substituted in the cubic law, returns the real value of transmissivity. The concept of hydraulic aperture is different from that of mechanical aperture, valid for plane fractures with regular apertures (par. 1.2.1).

There are many indirect techniques to estimate fracture aperture, based on laboratory measurements in stationary conditions and in situ tests. For laminar flow through parallel fractures, characterized by absence of roughness along the walls, the hydraulic aperture is defined by (Indraratna & Ranjith, 2001):

$$b_H = \left[\frac{12q\mu}{W \cdot i}\right]^{1/3} \tag{30}$$

where:

q = discharge flowing in stationary conditions
W = fracture width
i = hydraulic gradient
μ = fluid dynamic viscosity

Figure 1.24 shows experimental flow rate measurements obtained from a single fracture in claystone (Cook et al, 1990). The solid curve represents the flow rate through the fracture as a function of the aperture determined experimentally (hydraulic aperture); on the other side the broken curve represents the predicted flow rate, obtained by substituting the mechanical aperture b_M into the cubic law. When the fracture is propped open, predicted and experimental flow rates are quite close to each other; on the contrary as the fracture is closed, deviation of the experimental data from the cubic law increases rapidly; fracture permeability decreases more steeply than predicted by the cubic law (National Research Council, 1996).

Fracture roughness promotes frictional losses and reduces intrinsic permeability, as showed in figure 1.25 where the permeability obtained by using b_H=b_M (cubic law) is compared with the one calculated using the b_H obtained from Eq. [6]. The impact of roughness on permeability is greater for small apertures, resulting in a decrease of more than 4 orders of magnitude.

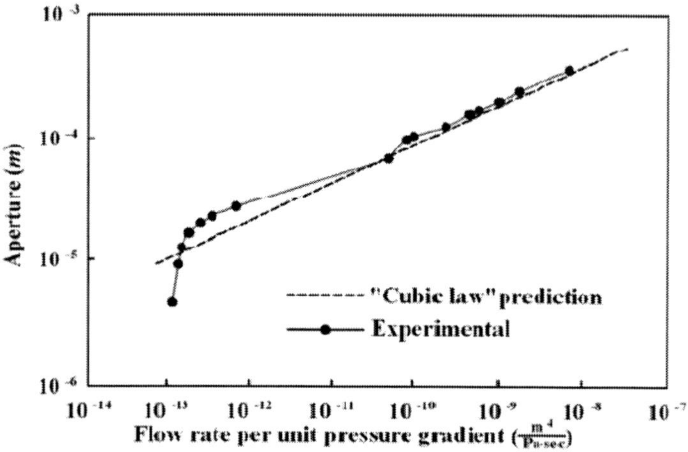

Figure 1.24. Deviation from the cubic law for partially closed fractures (National Research Council, 1996).

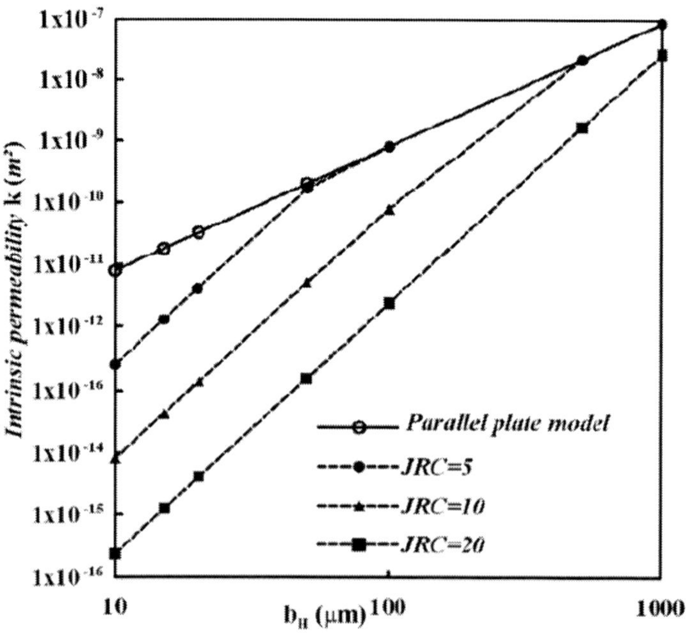

Figure 1.25. Intrinsic permeability as a function of fracture hydraulic aperture in the parallel plate model, $b_H = b_M$ (straight line) and in presence of roughness (dashed lines) (Price 2005).

For completely open fractures, the cubic law can be still valid when corrected by a *friction factor f*, to account for the roughness of the fracture surfaces, such that:

$$b = \left[\frac{12q\mu f}{W \cdot i}\right]^{1/3} \quad (31)$$

The parameter f has been calculated by numerous researchers in the past, among who Lomize (1951), Louis (1968) and Quadros (1982). Their results are summarized in the following expression:

$$f = \left[1 + m\left(\frac{R_f}{2b_H}\right)^{1.5}\right] \quad (32)$$

where m=17, m=8.8 and m=20.5 according to Lomize (1951), Louis (1968) and Quadros (1982) respectively. The parameter R_f has been defined by Louis (1968), as the difference between the highest "peak" and the lowest "valley" of the physical wall roughness.

When the walls are smooth, R_f=0 and therefore f=1.

Witherspoon et al (1980) carried out studies on discontinuities constituted by different lithological rock types (granite, basalt and marble), with fracture apertures from 250 μm down to 4 μm, and confining stresses up to 20 MPa. Eq [32] was found to be valid for $R_f/2b_H$>0.0065 and the determined friction factor was comprised in the range 1.04 -1.65.

It is possible then to define the relative roughness coefficient R_r (figure 1.26) as:

$$R_r = \frac{R_f}{D_h} \quad (33)$$

Where R_f represents the mean height of asperities in the fracture (some measured values of R_f are in the range 0.002 ÷ 0.01).

The hydraulic diameter in a plane fracture is given by:

$$D_h = \frac{4A}{p} \tag{34}$$

Where A is the cross sectional area where flow takes place and p indicates the external perimeter of the section. For an extended fracture D_h is equal to the double of its aperture ($D_h = 2b$).

Using then the formulation of expression of R_r it is possible to express hydraulic conductivity in a fracture characterized by rough and laminar flow:

$$K_f = \frac{\rho g b^2}{12\mu\left(1 + 8.8R_r^{1.5}\right)} \tag{35}$$

If in 2D, increases in roughness reduce the assumption of the parallel plate model, in 3D, this results in increasing also flow path tortuosity in the fracture plane. Non parallel unmated surfaces in a fracture deviate flow path laterally in the fracture plane (Price, 2005).

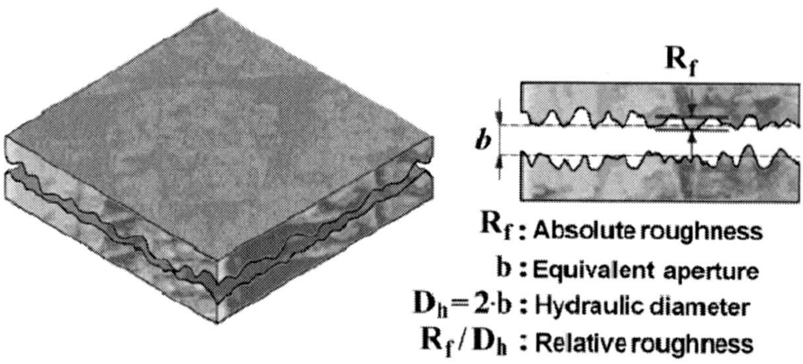

Figure 1.26. Roughness in a fracture.

Louis & Maini (1970) accounted for tortuosity in the expression of fracture hydraulic conductivity as a function of joint surface roughness and contact area such that:

$$K_f = \frac{\kappa g b^2}{12\nu f} \tag{36}$$

Where κ represents the open surface area of the joint divided by the total surface area and f is the friction correction factor determined from Louis (1969).

Walsh (1981) accounted for contacting asperities within an idealized fracture plane; he considered several circular inclusions in an idealized smooth walled parallel plate fracture and determined a reduction factor for permeability equal to (1+d)/(1-d), where d represents the proportion of contact area measured across the fracture surface.

In order to account also for narrowing of aperture, Cook et al (1990) included also the term f such that:

$$Q = \frac{1}{f}\left(\frac{1-d}{1+d}\right)\frac{b^3 \cdot w \cdot i}{12\mu} \tag{37}$$

Other than in presence of pronounced asperities, the cubic law loses validity also in cases of high hydraulic gradients.

In these circumstances, the laminar behavior is replaced by a transition one or even a turbulent one. In the case of turbulent flow, groundwater flow velocity can be expressed as: $V=K\cdot i^{\beta}$ where β is the degree of non linearity of flow (0.5<β<1). In fractured rocks it is difficult to determine the Reynolds number because for a given type of flow it can vary strongly from one point to another inside the same fracture, In this case it is convenient to express it in terms of D_h:

$$R_e = \frac{VD_h}{\nu} \tag{38}$$

Where V is the mean velocity, D_h is the hydraulic diameter of a fracture of aperture b and υ is the fluid kinematic viscosity.

The passage between laminar flow (β=1) to totally turbulent flow (β=0.5) is progressive, because the exponent varies slowly from 1 to 0.5 with the Reynolds number changing from 100 and 2300. Lee & Farmer (1993) have showed that the critical Reynolds number R_{ek} varies between 100 and 2300, and decreases with the increase of the relative roughness of the fracture. Therefore laminar flow takes place up to a critical Reynolds number (R_{ek}) equal to 2300 for hydraulically smooth surfaces and becomes turbulent with the increase of flow velocity and roughness. Turbulent flow conditions occur for values of R_{ek} starting from 100 in presence of high roughness.

The transition between laminar to turbulent flow takes place in correspondence of R_e equal to 2300, while that between smooth to rough flow in correspondence of R_r equal to 0.033.

On the basis of these considerations it can be affirmed that:

for $R_f/D_h \leq 0.33$ f=1 (39)

for $R_f/D_h > 0.33$ $f = \left[1 + m\left(\frac{R_f}{2b_H}\right)^{1.5}\right]$ (40)

Louis (1984) defined empirically five regimes of stationary flow, both laminar and turbulent, that depend of the various degrees of relative roughness, on fracture aperture, on the hydraulic gradient in the fracture and the kinematic viscosity.

1.5. Applicability of Cubic Law in Real Fractured Rock Aquifers

The cubic law that adapts Darcy law to flow through fractures is valid for ideal fractures characterized by flat, smooth and parallel walls, circumstance that doesn't happen in reality, being natural fractures characterized by variable surfaces and geometry. In sedimentary rocks at fine granulometry such as shales it is possible to find relatively smooth and flat fractures but in crystalline rocks as granite, fracture surfaces are rough and with variable apertures. Moreover, if subjected to stress, some fractures can be partially closed.

Witherspoon et al. (1980) affirm that the cubic law is valid for aperture up to 4 μm. According with the same author deviations from this law are detected when, due to local contact points strains are generated that overcome 20 MPa. An increase in the contact surface between fractures lowers the discharge that flows through them, even of two or more magnitude orders. The parallel plate model moreover is no more valid when channeling phenomena occur, frequent at high depths (Tsang & Tsang, 1987).

Besides, there are other complications connected with the applicability of Darcy law to fractured rock aquifers. The preferential orientation of fractures gives rise to anisotropy and the extreme variability of hydraulic conductivity even along the same direction is index also of heterogeneity.

As it has been demonstrated (par 1.3.2), the aquifer hydrodynamic parameters (hydraulic conductivity, porosity, dispersivity etc etc) can vary with the scale of the domain of investigation; moreover, even in cases in which a REV can be defined, it is not always possible to measure hydraulic conductivity at this scale. These circumstances cause complications connected to the interpretation and the subsequent prediction of flow and transport processes in fractured media.

When Darcy law is applied to fractured aquifers, it is important that hydraulic conductivity and hydraulic gradient are measured in the same portion of aquifer, intended both as test spatial localization and as depth of investigation.

The presence of heterogeneity can lead to obtain erroneous results when discharge is calculated multiplying regional hydraulic gradients by hydraulic conductivity values determined by means of pumping tests that operate at much lower spatial scales.

This concept is to be intended in terms of vertical and horizontal heterogeneity. With respect to the former, inside a single borehole, for example, zones characterized by higher conductivity values do not always correspond to higher flow zones, because the hydraulic gradient is not always the same for all depths (Cook, 2003).

As far as horizontal heterogeneity, problems can arise when piezometry is constructed starting from hydraulic head data sparse in a wide territory, being hydraulic conductivity highly variable also spatially at small scale.

In the site of Claire Valley (South Australia), the regional piezometry indicates an hydraulic gradient of about $3 \cdot 10^{-2}$ (60 m in 2 km) and the hydraulic conductivity K, measured between 35 and 38 m of depths, has proved to be 2 magnitude orders higher than the values measured at other depths. The use of these data returns a higher discharge than the real one to which a total recharge of 1640 mm/y corresponds, three times higher than the mean annual precipitation. Further investigations carried out have allowed to verify that the hydraulic gradient in that zone is lower than the regional one, probably because of the not good interconnectedness of fractures to the regional flow system at that depth (figure 1.27).

High degrees of fracturing detectable locally even at high depths may not correspond to equally high discharges because the fractures can be not interconnected to the regional flow system, like in the case of a well drilled in a low hydraulic conductivity zone that intercepts a highly fractured isolated zone.

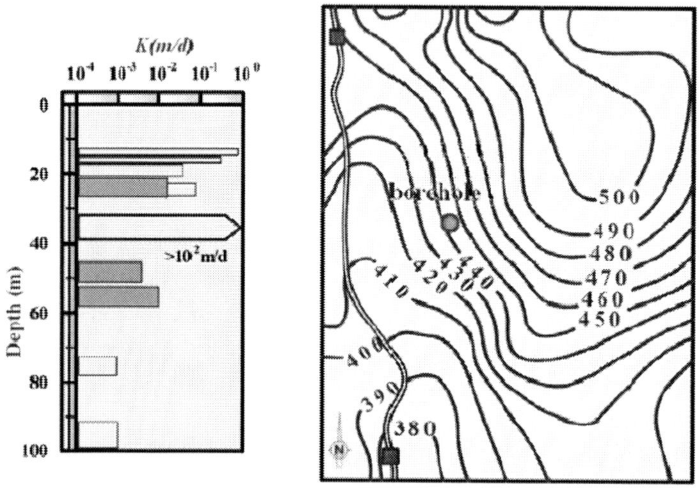

Figure 1.27. Vertical distribution of hydraulic conductivity and piezometry in the site of Clare Valley (Australia) (modified by Cook, 2003).

Measures of the variations of hydraulic head inside single piezometers can provide important information on the connectivity of fracture network (Cook, 2003). A distribution of hydraulic heads uniform with depth can be attributable to the presence of interconnected fractures. On the contrary, the presence of significant vertical variations in hydraulic head can suggest a low degree of connection between fractures. In figure 1.28 is represented the distribution of hydraulic head and conductivity along the vertical in a well in the Claire Valley zone.

A strong decrease in the head value between 35 and 40 m corresponds to a decrease in K, while at higher depths (60 m) the hydraulic head keeps the same values while hydraulic conductivity increases again: this is index of the fact that at this depth fractures are not interconnected to the regional flow system.

In relation to horizontal heterogeneity, hydraulic tests carried out in multiple wells at Mirror Lake (New Hampshire, USA), indicate that the bedrock contains a small number of highly transmissive fractures embedded within an extensive network of less transmissive fractures (Singhal, 2008). The highly transmissive fractures appear to form local clusters that are characterized by low connectivity at regional scale (figure 1.29). Each fracture cluster occupies a near-horizontal, tabular-shaped volume which has vertical thickness equal to 1.5 m approximately and horizontal extent between 20 and

50 m; the mean hydraulic conductivity is equal to $6{\cdot}10^{-5}$ m s^{-1}. Over distances of tens of meters, groundwater may flow through these highly permeable clusters, but over distances of 100 m or more, fluid flow takes place through less permeable regions which connect the clusters (Cook, 2003). The effective hydraulic conductivity at a scale of 100 m x 100 m has been estimated equal to be only $2{\cdot}10^{-7}$ m s^{-1} (Tiedeman et al., 1998).

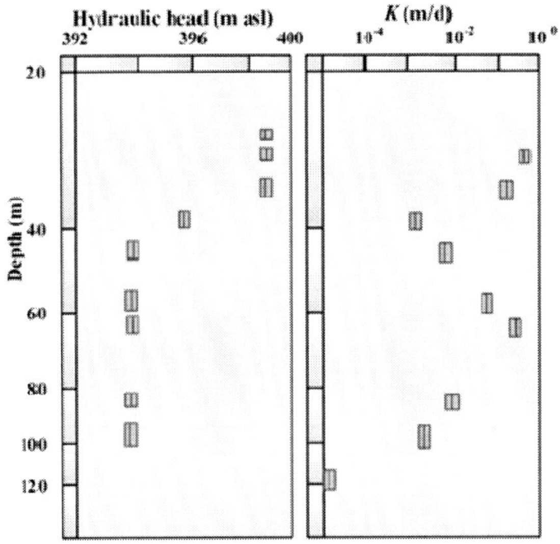

Figure 1.28. Vertical distribution of hydraulic head and conductivity in the Claire Valley zone (modified from Cook et al., 1999).

Therefore, if relevant vertical variations of hydraulic head exist, the construction of piezometric surfaces proves to be a difficult task. In analogy with porous media, in which piezometric surfaces are drawn for each aquifer, even in fractured media they should be constructed for each homogeneous fracture zone. Together with vertical heterogeneity, also the presence of horizontal heterogeneity and anisotropy create difficulties in drawing up piezometric surfaces. In fractured rock aquifers preferential orientation of fractures will usually result in a flow direction that is not perpendicular to the potentiometric contours. In theory, flow direction may differ from the direction of maximum hydraulic gradient by up to 90°; therefore potentiometric surface maps are not always able to be readily interpreted in terms of flow direction. High conductivity regions may result in preferential flow pathways that may be responsible for most of the flow. In fractured and karstic aquifers, a regional hydraulic gradient from recharge areas towards discharge areas justifies the drawing of regional contour maps, but the presence of local flow patterns and preferential pathways due to fractures, fault zones, karstic conduits and channels must be taken into account (figure 1.30).

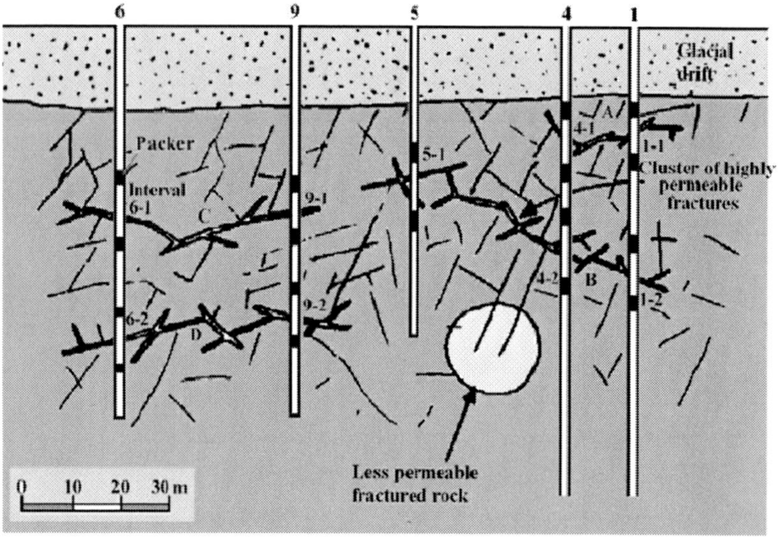

Figure 1.29. Vertical cross section of the US Geological Survey's fractured rock research at Mirror Lake, New Hampshire (modif from Shapiro & Hsieh, 1994).

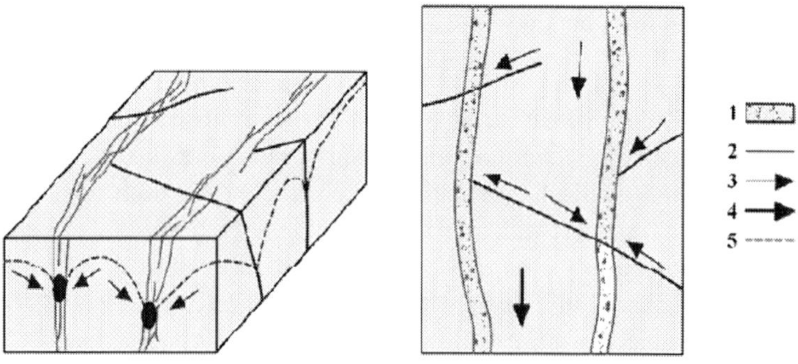

Figure 1.30. regional and local groundwater flow in a fractured and karstic aquifer. a) section view; b) plan view. (1) preferential flow pathways; (2) discontinuity; (3) local flow direction; (4) regional flow direction; (5) hydraulic head position (modif from Krešić, 1991).

In figure 1.31 it is shown the cone of drawdown after five days pumping in a fractured aquifer, in a well deep 100 m with a pumping rate of 20 ls^{-1}. The cone shows highly anisotropic characters in direction North East-South West (Caswell, 1992). In fact, the pumping induces a drawdown in the monitoring wells located 700 m north westly, while it doesn't show effects in a piezometer localized 150 m to the north.

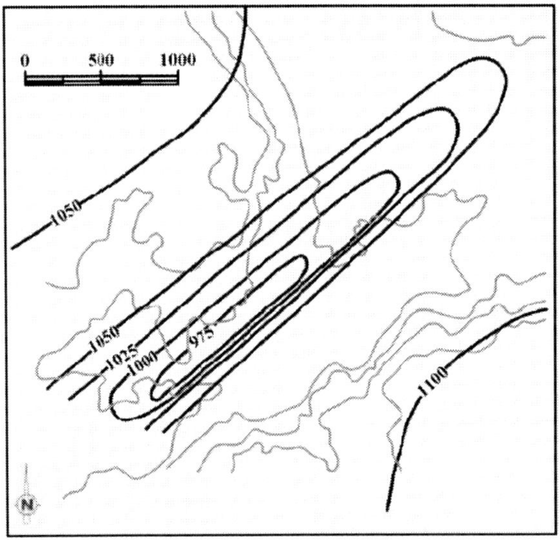

Figure 1.31. Anisotropy of cone of depression in a fractured aquifer (Caswell, 1992).

This concept is fundamental also when dealing with remediation treatments in fractured media. A contaminant plume that does not follow the hydraulic gradient may indicate that anisotropy is influencing the direction of groundwater flow. Figure 1.32 illustrates how horizontal anisotropy in fractured rock can affect the location of the capture zone of a pumping well. In the example, preferential orientation of fractures causes hydraulic conductivity to be higher in north-south rather than in east-west direction; therefore, the direction of groundwater flow will diverge from the direction of the hydraulic gradient (figure 1.32b). In this example, locating a pumping well on the basis only of the hydraulic gradient (figure 1.32a) would result in its failure to capture any portion of a contaminant plume, except in the immediate vicinity of the well (USEPA, 1996).

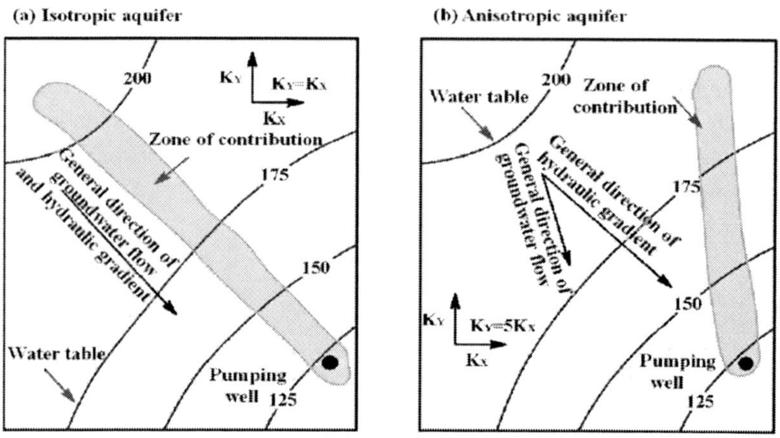

Figure 1.32. Effect of fracture anisotropy on the orientation of the zone of contribution (capture zone) to a pumping well (Bradbury et al., 1991).

In the case study by Brown (2005), the regional groundwater flow is from East – North East (figure 1.33a), while the plume is spread out southeasterly due to the presence of "dominant" fractures that create preferential pathways (figure 1.33 b). Therefore, a monitoring piezometer located in the direction of regional groundwater flow will fail to detect the presence of contamination.

If instead, pumping from a well causes greater drawdown in distant monitoring piezometers than in nearby ones, and they all lie on a straight line, so that the response cannot be explained by anisotropy, such a response indicates the presence of heterogeneity, that is to say local highly transmissive fracture zones (Hsieh, 1998).

Therefore, due to strong anisotropy and heterogeneity of fractured aquifers, it is a great advantage to monitor wells in various directions, distances (Nielsen, 2008) and depths in order to estimate the extent and degree of interconnection between fractures.

As far as temporal variation in hydraulic head in fractured rock systems, they can also be more pronounced than in porous media systems, due to the dual porosity-permeability behavior of such media. As will be explained in details in the next paragraphs, the prevailing of gravity effects in the fractures and of capillarity in the matrix will cause rapid recharge along discrete, permeable fractures, while the surrounding micropores in the matrix remain unsaturated. Also drainage will occur rapidly for the fractures while water will be held by capillarity in the finer pores in the matrix. As the time for the matrix to drain may be extremely slow, the aquifer matrix may supply water as baseflow to streams for many months after the water table has declined (Price et al., 2000). In figure 1.34 it is schematized the saturation and desaturation of a fractured rock. The initial water table position is indicated by the solid horizontal line. In case of rapid rising water table the matrix remains unsaturated (1), due to its low permeability, as the water level in the fracture rises. Instead, where the rate of water level rise is slow, and the matrix permeability is higher (3), the water table rises evenly in both fractures and matrix. More usually, an intermediate behavior occurs, in that the matrix will partially fill as the water table rises in fractures (2). For declining water table instead, the matrix will drain immediately (3) if gravity effects dominate on capillarity, or in the opposite case, remains saturated (1); in this case also an intermediate behavior will be occur (2) (Healy & Cook, 2002.)

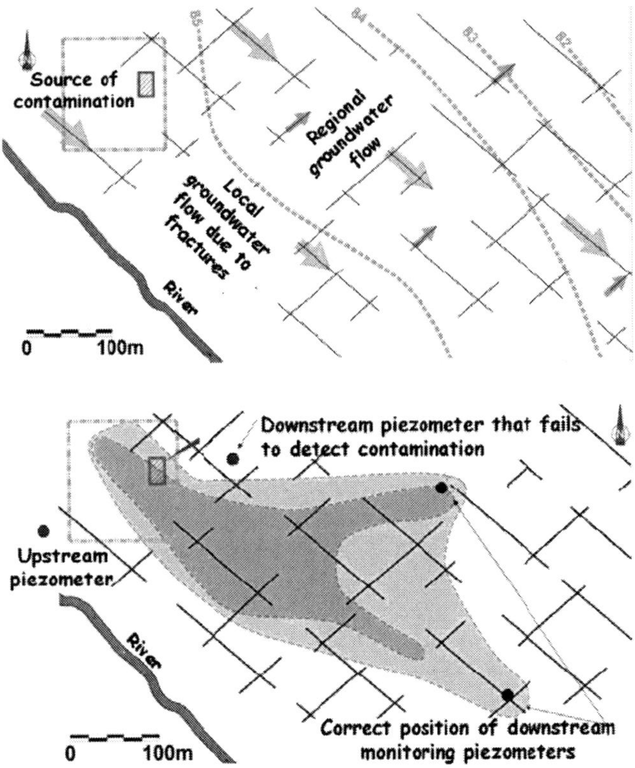

Figure 1.33. a) regional and local groundwater flow; b)anisotropy of plume propagation (modif from Brown, 2005).

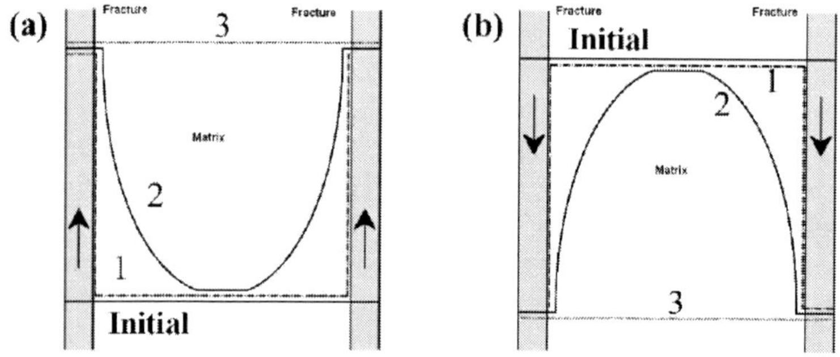

Figure 1.34. Schematic representation of saturation a) and desaturation b) of a fractured rock (Healy & Cook, 2002).

2. Unsaturated Fracture Flow

2.1. Moisture Content, Capillary Pressure and Hydraulic Conductivity for Fractured Media

The potential for migration of contaminants in fractured rock media depends on the rate of infiltration of moisture (Inyang& Bruell, 2000).

Groundwater circulation in unsaturated fractured rock masses is essentially a problem of multiphase fluid flow (water and gas phases) as in the unsaturated porous medium (Singhal & Gupta, 1999).

At the macroscopic scale the seepage of liquids through unsaturated porous media is usually described by Richards' equation [*Richards*, 1931], rewritten below in multiphase notation:

$$\frac{\partial(nS_l)}{\partial t} = \nabla \cdot \left[\frac{k_s k_r}{\mu_l} (\nabla P_l - \rho_l g) \right] \tag{41}$$

where k_s is the absolute (saturated) permeability of the porous medium, k_r is the relative permeability ($k_s \cdot k_r$= k actual permeability), ρ_l is the liquid density, P_l is the liquid phase pressure, g is the gravitational acceleration vector, n is the porosity, or volume of voids per volume of porous media, and S_l is the saturation of phase l, or fraction of void space occupied by phase l.

In partially saturated porous media, the relation between fluid pressure and saturation, which is described by the soil moisture characteristic curve, is governed by the effective capillary radii of the pores. During the drainage process, as the water phase pressure is decreased below atmospheric pressure, first there will be the desaturation of the largest pores and then of successively smaller pores.

Therefore, it is easy to infer that soil texture in a partially saturated medium strongly affects the curve shape. In sandy soils, where most of the pores are relatively large, once they are emptied, only a small amount of water is left; this corresponds to a steeply inclined curve. By contrast, clayey soils are characterized by smaller pores and therefore by a greater water retention; as a consequence there will be a more gradual decrease in wetness that corresponds to a more gradual slope of the soil moisture characteristic curve (figure 2.1).

When dealing with unsaturated media, the parameter effective permeability k is given as the product of the saturated permeability, k_s, (an intrinsic characteristic of the medium) and the relative permeability, k_r. The latter is a function of the unsaturated flow situation, and varies between 0 and 1 at residual saturation and full saturation, respectively. According with Gardner (1958), the relative permeability in unsaturated flow can be expressed as a simple exponential function of the capillary pressure using a single fitting parameter α, representing the relative rate of permeability decrease with increasing capillary pressure. The quasi- linear approximation for hydraulic conductivity in unsaturated porous media can be expressed by the following:

$$k = k_s \cdot k_r \quad k = k_s \cdot e^{\alpha\psi} \tag{42}$$

Where ψ is the moisture potential ($\psi = P_c/\rho g$) and P_c is the capillary pressure. Physically, the inverse of α can be interpreted as the air entry pressure, that is to say the minimum pressure needed to drive air into a water-saturated system.

Analogously with soil moisture characteristic curve, coarse-textured soils typically exhibit critical air- entry phenomena (smaller air entry pressure and a more abrupt variation of hydraulic conductivity with capillary pressure) compared with fine- textured soils.

Figure 2.2 shows the hydraulic conductivity vs. capillary pressure relationships for soils of different textures, i.e. for a fine-grained soil and a coarse-grained soil.

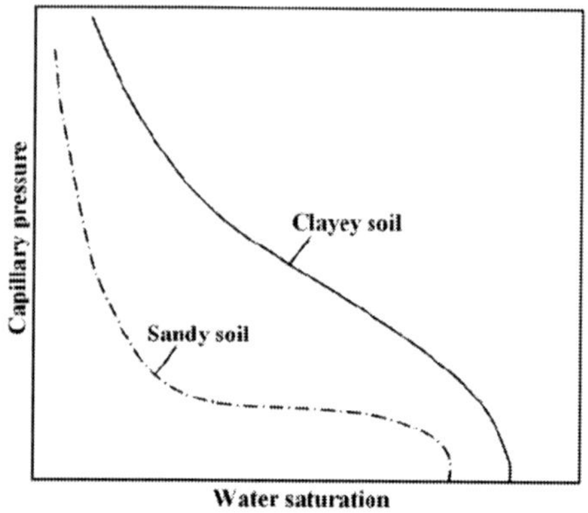

Figure 2.1. the effect of texture on soil- water retention.

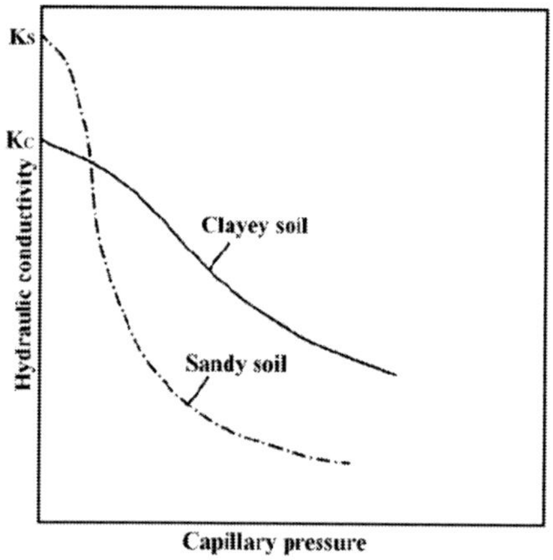

Figure 2.2. Hydraulic conductivity as a function of P_c for soils of different textures.

Although the saturated conductivity of a sandy soil K_s is normally higher than that of a clayey soil K_c, the unsaturated conductivity of the former decreases more steeply (due to the rapid drainage of large pores of a coarse-textured material) with increasing capillary pressure suction until it reaches negligible values, and eventually becomes lower. Consequently, in heterogeneous soils the unsaturated hydrodynamic behavior may be totally reversed: high-flux zones at near saturation become low-flux zones in highly unsaturated conditions and vice versa (Birkholzer & Tsang, 1997).

As far as heterogeneous soils are concerned, in fractured media fractures behave more similarly to sandy soils (1/α smaller, bigger pores) and matrix to clayey soils (1/α higher, smaller pores). It is therefore easy to infer that the fractures in a fractured porous medium will tend to remain dry under partial saturation and that water will be held by capillarity in the finer pores of the matrix (Singhal & Gupta, 1999).

In heterogeneous fractures the preference of the infiltrating liquid for large or small apertures is determined by the combined effects of gravity and capillary forces. The ratio between gravity and capillary forces is expressed using the Bond number (B_o):

$$B_o = \frac{\text{gravity force}}{\text{capillary force}} = \frac{\Delta\rho b^2 g \sin\beta}{\sigma \cos\gamma} \tag{43}$$

where $\Delta\rho$ is the density difference between the infiltrating liquid and the air, g is the gravitational acceleration constant, β is the angle of inclination of the fracture measured from the horizontal, b is the aperture, σ is the liquid surface tension, μ_l is the viscosity of the liquid, u_l is the velocity of the liquid, and γ is the contact angle of the wetting phase on the solid surface. Larger gravity forces (high Bond numbers) will drive the denser fluid into the larger apertures.

The ratio between viscous and capillary forces is the capillary number (C_a):

$$C_a = \frac{\text{viscous force}}{\text{capillary force}} = \frac{\mu_l u_l}{\sigma \cos\gamma} \tag{44}$$

By contrast, larger capillary forces (small capillary numbers) will drive the wetting phase into the smaller apertures.

Capillary pressure determines (Wang & Narasimhan, 1985) the water-phase pressure at which a given size fracture or pore will drain. From force balance considerations, the height of rise for an idealized parallel-plate, smooth- wall fracture with aperture b is:

$$h_e = -h = \frac{2\gamma\cos\vartheta}{b\rho g} \tag{45}$$

With gravitational acceleration g, water density ρ (=1000 kg/m^3), surface tension γ (=0.07183 kg/s^2), and angle θ (=0) between liquid surface and solid surface remaining constant at ambient temperature, the capillary pressure is inversely proportional to fracture aperture b. Using the capillary theory for a set of parallel, smooth-wall fractures at given negative pressure head, -h, fractures with apertures $b<b_c$ are assumed to contain wetting phase, whereas fractures with apertures $b>b_c$ will be drained.

Few studies have analyzed how best to average the properties of fracture-matrix systems in unsaturated conditions. A typical aspect of the soil moisture characteristic curve for fractures and matrix is reproduced in figure 2.3a: a steeply inclined curve for the fractures that corresponds to a faster drainage process (similarly to sandy soils) and a more gradual slope for the matrix that corresponds to a more gradual decrease in wetness due to higher retention (similarly to clayey soils).

Few studies have examined how to represent and average the properties of fractures and matrix under partially saturated conditions. Montazer and Wilson (1984) introduced a relation between effective permeability and capillary pressure for a combined fracture-matrix system.

As represented in figure 2.3b, the combined uppermost parts of a fracture curve and a matrix curve can be combined to represent an "equivalent continuum" (NRC, 2001). From this representation it can be inferred that at very low levels of saturation matrix capillary effects dominate, and high saturation levels are required for fractures to conduct water.

Also Peters and Klavetter (1988) defined a "*composite conductivity*" able to average the properties of fractures and matrix at near and full saturation (figure 2.3c). Being both relative (k_r) and intrinsic conductivities (k) a combination of the fracture and matrix relative conductivities (Ho, 1997), the resulting volume-averaged product of the intrinsic permeability and the relative permeability can be expressed as follows [Peters & Klavetter, 1988]:

$$kk_r = k_f k_{r,f} \phi_f + k_m k_{r,m} (1 - \phi_f) \tag{46}$$

where the subscripts *f* and *m* denote the fracture and matrix materials. The fracture and matrix saturations are then used in the following equation to determine the effective continuum saturation [Peters and Klavetter, 1988]:

$$S = \frac{S_f \phi_f + S_m (1 - \phi_f) \phi_m}{\phi_f + (1 - \phi_f) \phi_m} \tag{47}$$

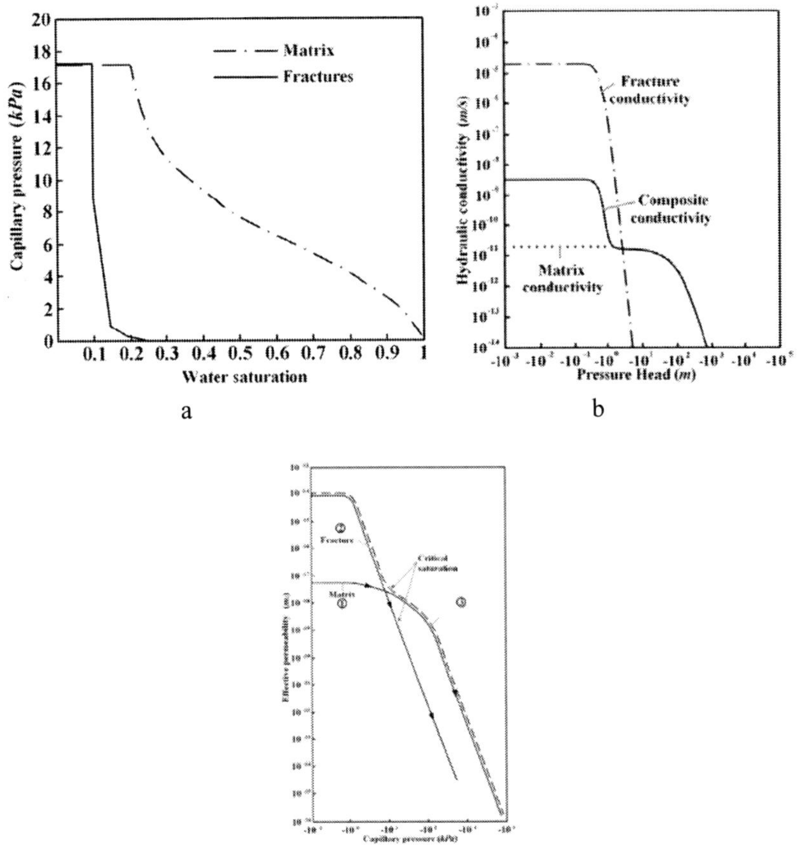

Figure 2.3. a: capillary pressure/Pc for fracture and matrix; b) effective permeability/Pc for fracture and matrix (Montazer & Wilson, 1984); c) Hydraulic conductivity of unsaturated fractured volcanic tuff from Yucca Mountain, Nevada, (Peters & Klavetter, 1988) plotted as a function of pressure head. Pressure values are negative, indicating that water is held in tension.

Figures 2.4, 2.5, 2.6, 2.7 show the hydraulic conductivity as function of capillary pressure, and capillary pressure as function of saturation for 0.01, 0.1 mm fractures and the matrix at a fractured site (Cherubini et al, 2008).

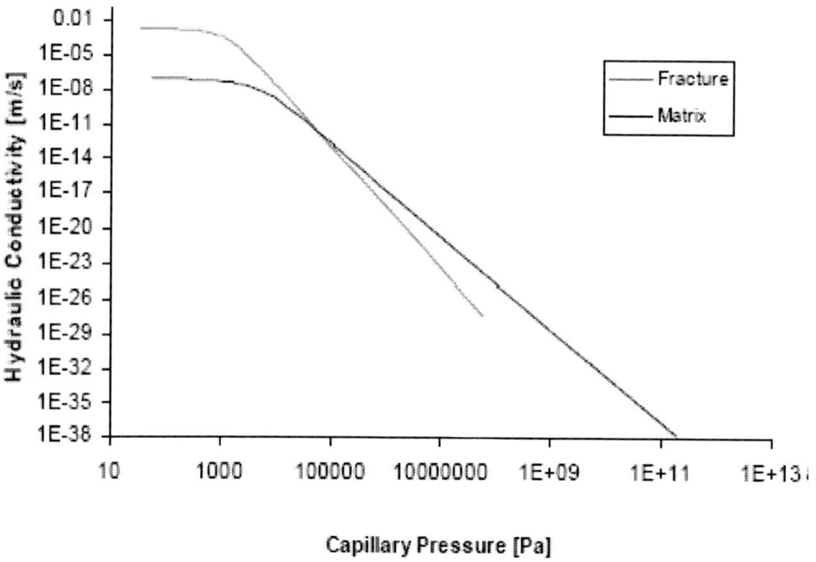

Figure 2.4. k/P_c curves for fractures (b=0.1 mm) and matrix.

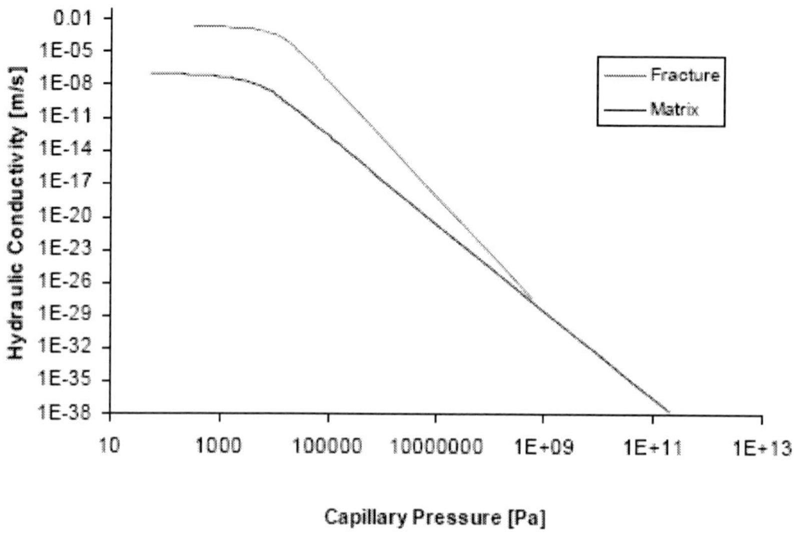

Figure 2.5. P_c/S curves for fractures (b=0.1 mm) and matrix.

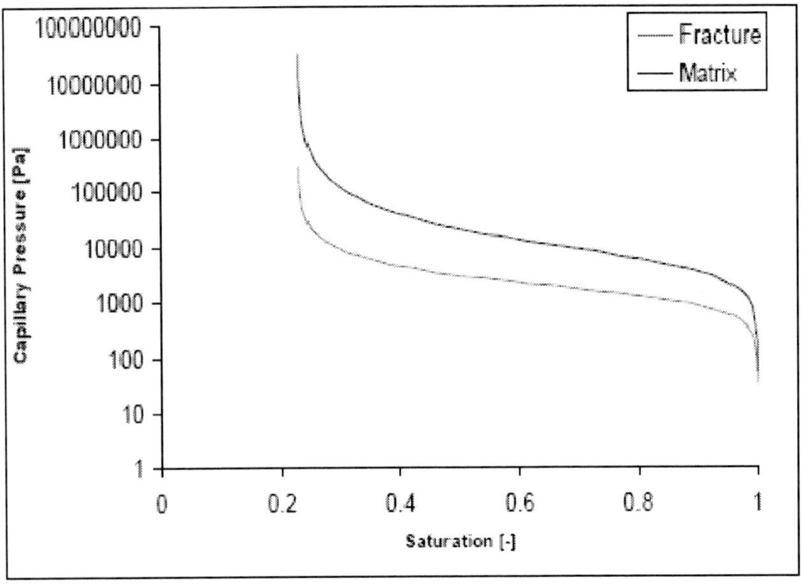

Figure 2.6. k/ P_c curves for fractures (b=0.01 mm) and matrix.

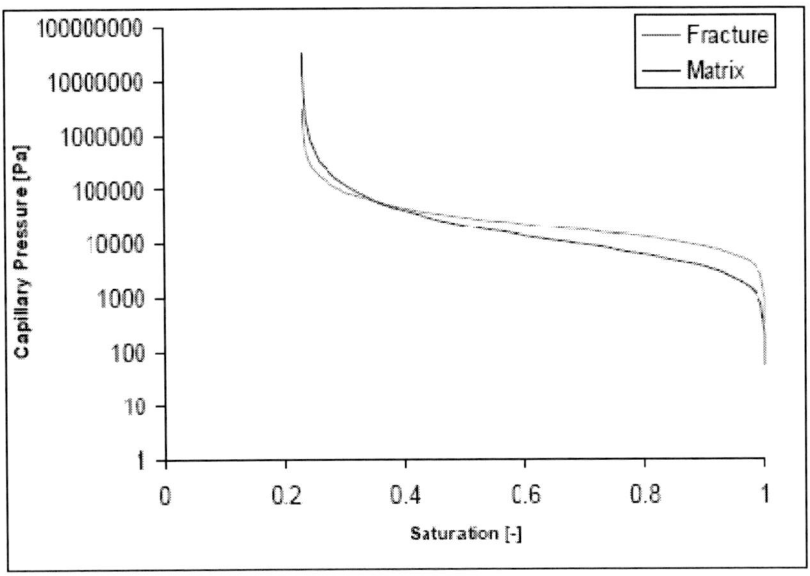

Figure 2.7. P_c/S curves for fractures (b=0.01 mm) and matrix.

From the capillary pressure/saturation curve it could be seen that at equal values of suction, the matrix has got higher saturation in case of fractures of significant apertures (b=0.1 mm; $1/\alpha$=1440). In other words, fracture flow appears to be insignificant as compared to matrix in unsaturated conditions. On the contrary, for very small fracture apertures (b=0.01 mm), the value of $1/\alpha$ rises up to 14400 and, as it strongly conditions the curve, it happens that at equal values of P_c, the fracture starts inverting the trend and has got higher saturation. The smaller the fracture is, the closer it is to clay, that is to say it has got higher capillarity. The same could be inferred by the hydraulic conductivities/capillary pressure curves. In case of higher fracture apertures (b=0.1 mm) for low values of capillary pressure the fracture is more conductive, whereas as the moisture content decreases (P_c increases) matrix flow starts dominating. Instead, when fractures apertures are very small, like voids, they are always more conductive than the matrix.

From this analysis it is easy to infer that fractures will control the fluid flow at fully saturated conditions, while under unsaturated conditions, the role of fracture flow is minimal and the flow is mainly governed by matrix-block characteristics (Singhal & Gupta, 1999). Under unsaturated conditions, as fractures apertures are usually larger than matrix pores, the former will drain easily and will remain dry while the water will be held by capillarity in the finer pores of the matrix (Wang & Narasimhan, 1985).

2.2. Fracture Hydraulic Properties

In a fractured medium it is necessary to apply more complex relative permeability/capillary pressure models which account for characteristic properties of the medium itself; for example, the pore size distribution coefficient *n*, being a measure of the spatial variability of pore sizes in a soil, in fractured rock, denotes the spatial variability of fracture apertures.

From capillary theory, the α parameter of a single fracture is correlated with fracture aperture and permeability, in that large apertures are typically associated with smaller capillary suction and higher permeability. The α value for an individual fracture can be approximately related to the corresponding average aperture by [Wang & Narasimhan, 1993]:

$$\alpha = \frac{b}{2\sigma \cos\theta} \tag{48}$$

where σ is the surface tension (0.07) and θ is the contact angle between liquid and solid surface.

Analogously with the Bond number, the parameter α, called the sorptive number, ranges from 10^{-1} to 10^{2} m^{-1} for natural soils (Pullan, 1990) and indicates the relative importance of gravity and capillary forces: if α is large, gravity effects dominate, as for fractures; if α is small, capillary forces dominate, as for the matrix.

Richards' equation has been applied to fractures, based upon the conceptual model that views a fracture as a twodimensional porous medium. The fracture is volume averaged over a large scale and modeled as smooth parallel plates [Peters & Klavetter, 1988]. This approach predicts the liquid will proceed with a spatially uniform infiltration front, subject to strong capillary imbibition effects from the rock matrix that draw the flowing liquid from the fracture [Nitao & Buscheck, 1991; Wang & Narasimhan, 1985].

According with Liu and Bovardsson [2001] the average aperture of a fracture, b, can be related to its corresponding trace length by:

$$b = cL^{d} \tag{49}$$

where c and d are empirical constants, and L is the trace length (m).

In reality, rock fractures have rough walls and variable apertures with numerous asperity contacts.

Where porous blocks are separated by desaturated fractures, there may not be any flow of water from one porous block to another.

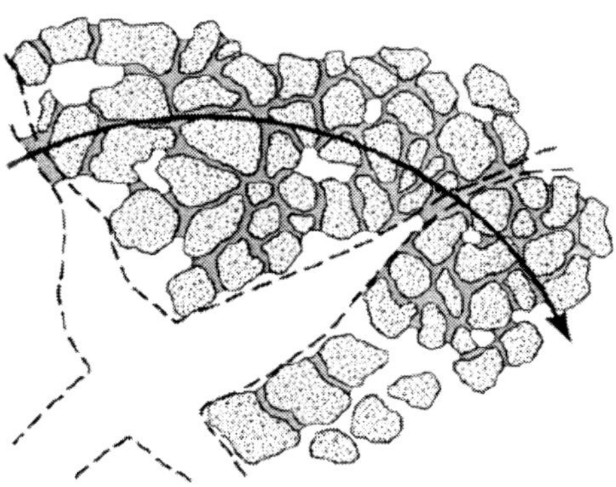

Figure 2.8. Conceptual model of water flow in partially saturated, fractured rock showing schematically the flow lines moving around the dry portions of the fractures. (Wang & Narasimhan, 1985).

However, at higher depth due to compressional stresses, the walls of the fractures are pressed together forming asperities that lead to partial closure of fractures. The reduction in joint aperture and therefore in permeability has an important effect in making fractures act as barriers to liquid flow: water will tend to flow from one matrix block to another across the fracture sections that remain saturated (figure 2.8) (Wang & Narasimhan, 1985). As dry portions of the fractures will be circumvented by flow lines, the result will be a more tortuous flow path, that is to say a macroscopic tortuosity introduced in the system.

3. Contaminant Propagation in Fractured Rock Aquifers

3.1. Mechanisms of Solute Transport in Fractured Rock

The issue of pollutant transport in fractured aquifers has been subject of research for several decades. This topic involves both a macroscopic description able to predict the propagation of contamination that mainly takes place throughout fractures acting as major conduits, but also a microscopic (or local) description that involves the mass transfer processes between fractures and matrix. This local scale study of solute transport is considered superior to the approach using macroscale because it takes into account the local advective-dispersive-diffusive properties behind the phenomena (Bai et al. 1995).

The contaminant propagation in fractured rock aquifers takes place according to the following processes:

3.1.1. Conservative Transport

- Advection at the mean fluid velocity that takes place mainly through the fractures. The advective component in the matrix is negligible due to the high contrast of permeability between fractures and matrix. Advection in the matrix can be relevant under significant local flux;
- Hydrodynamic dispersion in the fractures due to variations of local fluid velocities with respect to the mean velocity;

- Molecular diffusion in the fracture plane and from the fractures into the matrix (Abdalla, 2005).

The mechanisms that govern conservative solute transport depend on the relative porosities and permeabilities of the fractures and the matrix blocks

1) If the rock matrix is impermeable, has negligible porosity and is compact, advective transport through the fractures will be prominent.
2) If the matrix blocks are porous but have negligible permeability, solute transport will be characterised by advection through the fractures, with diffusion into the immobile water in the matrix; transport through the rock matrix by advection will be generally insignificant because of its low permeability: In such a case, as part of a cleanup operation, the contaminants from the fractures may be flushed out relatively quickly but the porous blocks will only slowly release the contaminants into the fractures.
3) In a third situation, when the matrix has the same permeability as the fractures, the transport of solute will take place simultaneously in the two media by dispersion and diffusion, depending on differences in solute concentration in the matrix blocks and the fractures.

3.1.2. Reactive Transport

- Physico-chemical reactions between solute, matrix and fracture walls (including geochemical processes such as acid-base reactions, oxide reduction reactions, precipitation/sorption/desorption reactions) that cause retardation or slowing of apparent velocities.
- Biotic or microbial-mediated transformation (Abdalla, 2005).

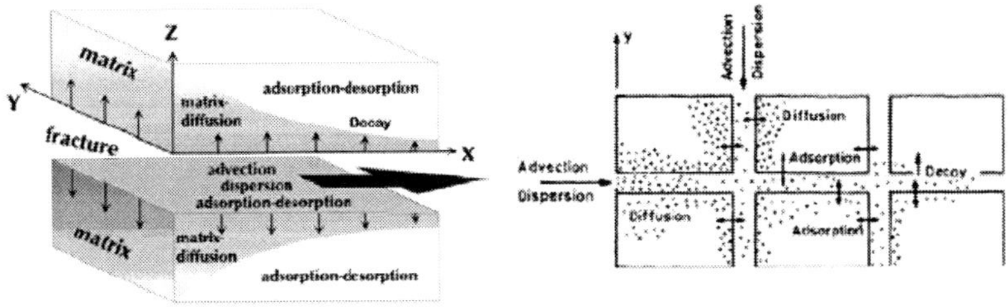

Figure 3.1. Schematic representations of various mechanisms of solute transport in an idealized fractured porous medium (Witthüse, 2001; Germain & Frind, 1984).

The solution of mass transport in fractured rock aquifers is governed by the following equations. It is assumed that the matrix is completely homogenous and isotropic and it extends infinitely away from the fracture (Tang et al., 1981; Lever et al.,1985). The solute transport in a fracture is:

$$\underbrace{R_f \frac{\partial C_f}{\partial t}}_{\text{Retardation}} + \underbrace{v_f \frac{\partial C_f}{\partial x_i}}_{\text{Advection}} - \underbrace{D_{h\,ij} \frac{\partial^2 C_f}{\partial x_i^2}}_{\text{Dispersion}} - \underbrace{\frac{2D_e}{b} \cdot \frac{\partial C_f}{\partial z}}_{\text{Diffusion}} \Bigg|_{z=\pm b/2} + \underbrace{\lambda R_f C_f}_{\text{Decay}} = 0 \tag{50}$$

Solute transport in the rock matrix (neglecting the advective component) is described by:

$$\underbrace{R_m \frac{\partial C_m}{\partial t}}_{\text{Retardation}} - \underbrace{\frac{D_e}{n_m} \frac{\partial^2 C_m}{\partial z^2}}_{\text{Diffusion}} + \underbrace{\lambda R_m C_m}_{\text{Decay}} = 0 \tag{51}$$

R_f retardation coefficient of fracture surface [-]
R_m retardation coefficient of rock matrix [-]
C_f concentration in the mobile (fracture) zone [ML^{-3}]
C_m concentration in the immobile (matrix) zone [ML^{-3}]
v_f average solute velocity in the fracture [LT^{-1}]
D_h hydrodynamic dispersion coefficient [L^2T^{-1}]
D_e effective diffusion coefficient [L^2T^{-1}]
i,j spatial coordinate direction x and z [-]
x spatial coordinate taken to be positive in the direction of flow in the fracture [-]
z direction of transport in the matrix perpendicular to the fracture [-]
n_m matrix porosity [-]
λ first order decay rate [T^{-1}]

Some tracer tests conducted in fractured rocks showed skewed breakthrough curves: early initial arrival times, with one or more sudden jumps in concentration, and/or long tails. The behaviour of the early-time (peak) is dominated by advection and dispersion: some of the tracer moves quickly through open channels resulting in an early peak, while the other portion is delayed by diffusion of tracer into the matrix and/or into stagnant water zones in the fracture resulting into a tail with low concentration (Abdalla, 2005). The slope of the tail contains information about the type of mass transfer (e.g. single-rate or multi-rate diffusion or first-order sorption)

(Haggerty, 2000), whose rate depends essentially on the differences in concentrations and permeabilities of the two interacting media.

Figure 3.2 shows a breakthrough curve obtained in a fractured medium, whose concentration change at each monitoring location shows a reduction compared with the expected results. Figure 3.2a can be schematically explained by figure 3.2b, where concentration changes in time actually replicate the observed breakthrough curve.

This phenomenon can be attributed to different processes, such as micropore-macropore interaction, tortuosity, pore clogging, and finally to the attenuation effect of molecular diffusion in the matrix.

In fact slow and diffusive flow phenomena occurring within the matrix can significantly modify the flow rate in the primary flow channels, the fractures (Bai et al, 1995); these processes may cause solute attenuation or, conversely, may give rise to replenishment depending on the relative concentration between matrix and fractures.

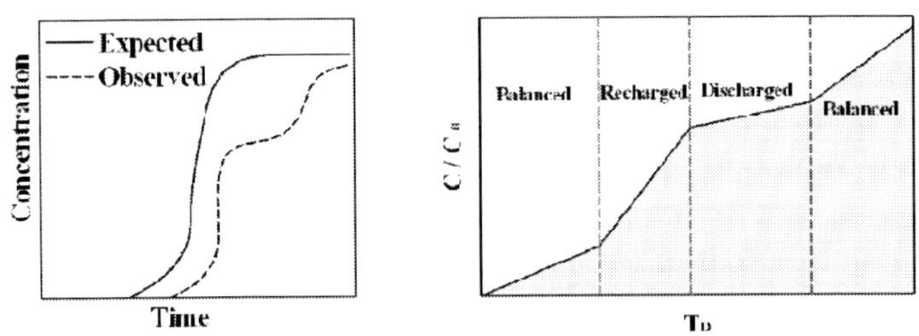

Figure 3.2. a) Expected and measured solute breakthrough in a fractured medium b) schematic solute breakthrough curve (Bai et al., 2000).

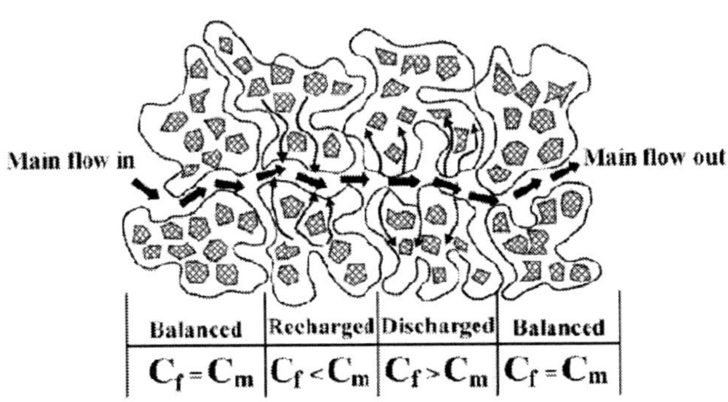

Figure 3.3. mechanism of mass exchange between fractures and matrix (Bai et al., 2000).

It is therefore necessary to perform an accurate characterization of flow in matrix blocks. As shown in figure 3.3, when $C_f = C_m$ the contaminant is primarily transported through fractures. The rate of transport in fractures begins to increase if the concentration in fractures is less than in the matrix, that is to say $C_f < C_m$; this process has been called matrix replenishment (Bai et al., 2000), and in this circumstance fractures become a sink. Replenishment results in additional solute spreading, both in primary flow channels (figure 3.3) and in secondary flow pathways (Bai et al., 1995). The conceptualization of the matrix replenishment phenomenon can be referred to the papers by Bai & Roegiers (1995, 1997), Bai et al. (1996). Conversely, there is a reduction in the rate of transport in the fractures if the concentration in the matrix is less than in the fractures, that is to say $C_f > C_m$. This is the process of matrix diffusion that attenuates pollutants migration along the fractures and will be explained in detail in the following paragraphs. In this circumstance fractures become a source.

Numerous tracer tests (e.g. Becker & Shapiro 2000; Einsiedl et al. 2000) proved that the tailing effect of breakthrough curves is not always a result of matrix diffusion processes (Abdalla, 2005). Becker & Shapiro (2000) showed in tracer tests in a fractured crystalline rock that tracers with different diffusive properties produced identical late time breakthrough. This result implies that the tailing was not caused by any purely diffusive transport mechanism rather it resulted from hydrodynamic transport mechanisms such as advection or hydrodynamic dispersion due to highly heterogeneous spatial and temporal nature of transport in fractured rock (Abdalla, 2005).

Bouhroum & Civan (1994) showed that the extended tailing in breakthrough curves obtained by conducting laboratory experiments for fractured media was the result of velocity fluctuations between the high velocity area in preferential flow pathways and the low permeability zones. The skewness observed in the breakthrough curve may be therefore partially attributed to the impact of matrix advective term as a result of local velocity variations (Bai et al., 2000).

As a result of this analysis, it is of crucial importance to determine the initial concentrations as well as the properties of fractures and matrix in order to predict accurately the direction of contaminant transport (Bai et al., 2000).

3.2. Conservative Solute Transport

3.2.1. Advection

The advective component in the equation of solute transport in fractured rocks is quantified by means of the cubic law that combines Darcy equation with fracture hydraulic conductivity:

$$q = \frac{Q}{b \cdot w} = \frac{b^2 \rho g}{12\mu} \cdot i \tag{52}$$

Where q is Darcy's specific discharge, Q is volumetric flow through the aperture, i the hydraulic gradient and the term $b^2\rho g/12\mu$ represents the hydraulic conductivity.

The phase average velocity v is expressed by:

$$v = \frac{q}{n_e} = \frac{b^2 \rho g}{12\mu} \cdot \frac{i}{n_e} \tag{53}$$

Where n_e is the effective porosity and $n_e < n$ (n is the total porosity).

By computing groundwater velocities by using the modified Darcy's equation it can be showed that fractured rocks have much lower effective porosity (10^{-2} to 10^{-5}) compared to porous granular material (0.5 to 0.03). Therefore, groundwater velocities in fractured rocks are several orders of magnitude higher than in granular soils.

If we assume that a fractured medium (granite) has a fracture porosity (n_f) of 10^{-4}, and hydraulic conductivity of 10^{-5} ms^{-1} and if the hydraulic gradient is 10^{-2}, as is generally the case in the field, the groundwater velocity using this equation will be estimated to be equal to 86.4 $m \cdot d^{-1}$. This compared to a porous medium (say silty sand) with the same values of hydraulic conductivity and hydraulic gradient and effective porosity of 0.26, will have a velocity of about 0.03 $m \cdot d^{-1}$ indicating thereby that the groundwater velocity in a fractured medium can be several times higher than in the porous rocks. Therefore, advective transport is much faster in fractured media than in porous rocks provided the matrix is impermeable and has negligible porosity. However, numerical studies show that tracer velocities may be less or greater than the average fluid velocity depending on the fracture roughness and contact area. The tracer velocity is found to be less than the average fluid velocity if flow is transverse to the roughness, and greater if flow is parallel to the roughness (National Research Council, 1996).

3.2.2. Hydrodynamic Dispersion

In addition to advection, the pollutant has a tendency to spread by hydrodynamic dispersion. This process causes the dilution of solutes, spreading them both vertically and horizontally as they travel down gradient.

In fractured aquifers, dispersion is due to heterogeneities of the velocities in a fracture. During flow through a cylindrical pipe, the velocities are distributed in a parabolic form causing a faster movement of solute along the axis and a comparatively slower movement near the surface due to drag effect (figure 3.4) (Singhal & Gupta, 1999). Dispersion may be also due to differences in velocities from one fracture to another, because of variations in aperture sizes, fracture shapes and orientations. It may also be affected by the tortuosity and connectivity of multiple flow paths: intersection and channeling of fractures cause transverse mixing and dispersion.

These phenomena can, in some circumstances, invalidate the cubic law, in that the solute residence time proves to be higher than the one calculated with the smooth and parallel surface model. Dispersion is a non- steady irreversible process.

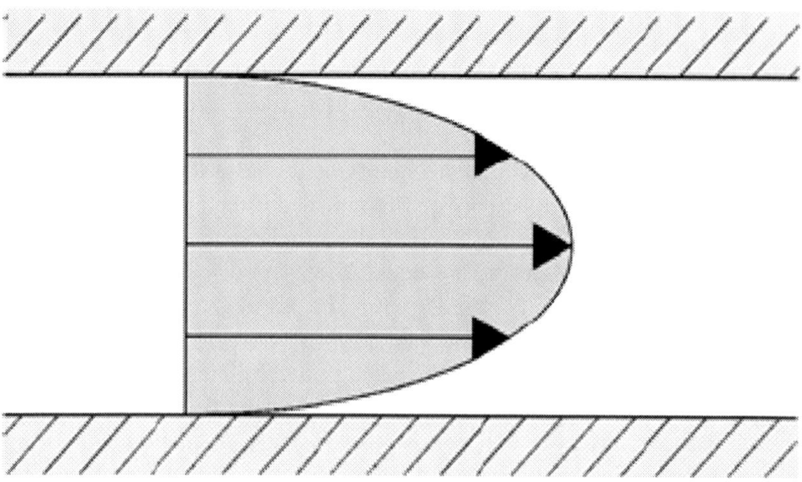

Figure 3.4. Distribution of velocities due to hydrodynamic dispersion through a pipe.

The hydrodynamic dispersion coefficient D_h describes both the mechanical D_d (dispersion due to the variation of the fluid velocity across the fracture) and the diffusive D_m mixing of the solute. For one-dimensional flow hydrodynamic dispersion is defined as (Bear 1972):

$$D_{hl} = \alpha_l \cdot V_x + D_m \tag{54}$$

$$D_{ht} = \alpha_t \cdot V_x + D_m \tag{55}$$

D_{hl} longitudinal hydrodynamic dispersion coefficient $[L^2T^{-1}]$

D_{ht} transverse hydrodynamic dispersion coefficient $[L^2T^{-1}]$

α_l, α_t longitudinal, transversal dispersivity [L]; characteristic properties of the medium

D_m coefficient of molecular diffusion $[L^2T^{-1}]$

At low velocity, the most relevant phenomenon is diffusion: the coefficient of hydro-dynamic dispersion, D_{hl}, equals the diffusion coefficient, D_m. At high velocity, mechanical mixing will be the dominant mechanism of solute transport, therefore $D_{hl} = \alpha_l \cdot V$ (Singhal & Gupta, 1999). The ratio of longitudinal to transverse dispersivity in an aquifer influences the shape and propagation of the contaminant plume.

Laboratory values of dispersivities will not be of much help in predicting dispersion in the field, as the are significantly smaller than those obtained from field studies (Freeze & Cherry, 1979) which is attributed to the effect of heterogeneities on field scale (Fetter, 1993). Methods of determining dispersivity in the field are given by Freeze and Cherry (1979). Field studies indicate that longitudinal dispersivity, α_l typically ranges from 10 to 100 m and transverse dispersivity, α_t is 10 to 30% of α_l (table 1). The fractured basalt of Idaho shows an exception, in that it has a greater value of α_t as compared with α_l which is due to fracturing (Singhal & Gupta, 1999).

The dispersivity within a single fracture is usually small and dependent on aperture, roughness and type and frequency of bridging materials within the fracture (Grisak & Pickens 1980). In a model where the apertures are correlated and lognormally distributed, Moreno et al (1988) found a strong dependence of dispersivity on the variance of the aperture distribution.

Dispersivity in a fractured medium (figure 3.5) is scale dependent; it increases with the scale of measurement, because of the heterogeneous nature of the aquifer materials (Abdalla, 2005).

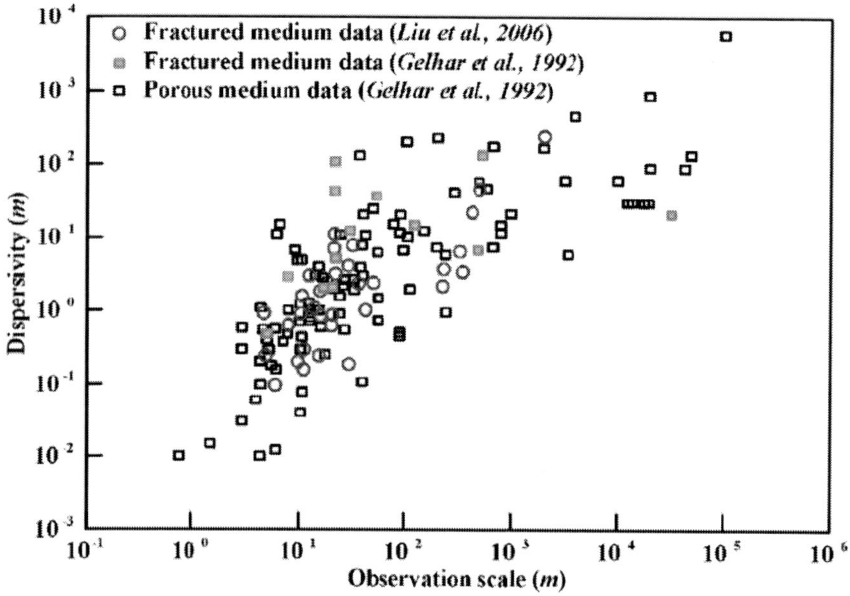

Figure 3.5. Longitudinal dispersivities as a function of scale of observation for different aquifer materials (Gelhar et al. 1992).

Table 3.1. Dispersivity values in different rock masses estimated from field studies (Fetter, 1988; De Marsily, 1986)

Rock	*Location*	*Dispersivity (m)*
Basalt	Hanford, Washington, USA	$\alpha_L = 30$ $\alpha_T = 20$
Basalt	Snake River, Idaho, USA	$\alpha_L = 91$ $\alpha_T = 137$
Limestone	–	$\alpha_L = 152$
Alluvium	–	$\alpha_L = 10–60$ $\alpha_T = 5–50$

Flow channelling within the fracture is widely recognized to have a large effect on dispersion (Berkowitz, 2002; Moreno & Tsang, 1994). Velocity variations in the fluid in a flow channel and velocity variations between the different flow channels in the porous media might be responsible for this effect (Abelin et al., 1991).

3.2.3. Channeling Phenomenon

Another phenomenon that has a substantial effect on solute transport in fractured media is flow channeling (Abelin et al., 1991; Tsang & Neretnieks, 1998; Tsang et al., 1988; Tsang, 1993).

In the fracture plane, the presence of asperities acts as restriction of the aperture, with the consequence that flow is focused along a few preferred pathways, that have the lowest hydraulic resistance (figure 3.6) (Tsang & Tsang, 1989) and have been called “flow channels”. These channels might transport a portion of the solute considerably faster than the average.

Within the same fracture there could be different channels that may not meet and mix their water over considerable distances, therefore solute transport can be considered separately for each channel (Abdalla, 2005).

Experimental and theoretical work has shown that flow channelling may be observed in either a single fracture or in a fracture network, and that it occurs on all scales, from centimeter to kilometer scale (Margolin et al., 1998; Moreno & Tsang, 1994; Tsang & Neretnieks, 1998).

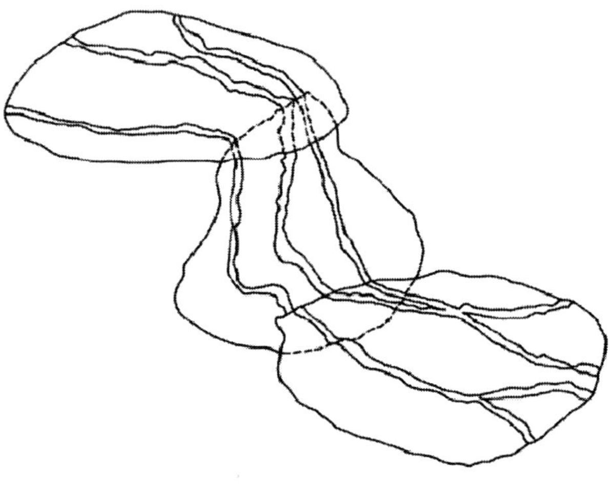

Figure 3.6. The channelling concept within the fracture (Tsang, 1993).

The effect of roughness of fracture walls on the transport of solutes is demonstrated in a study of Raven et al (1988) which revealed that transport through fractures characterized by wall irregularities creates not only the phenomenon of "flow channeling" but also of stagnant water zones between the channels within the fracture, and along the edge of the fracture (figure 3.7). However, the largest stagnant water body, by far, is that in the porous matrix of the rock (Tsang & Neretnieks, 1998).

This coupled channeling/diffusion process explains the behavior of breakthrough curves for fractured aquifers: during the initial time some of the solute moves quickly through channels resulting in a early peak, while the other portion is delayed by diffusion into the matrix and/or into stagnant water zones in the fractures resulting in a low concentration tail (Abdalla, 2005)

Tsang (1991) demonstrated that the rock surface available for surface sorption and diffusion might be only 10 or 20 % of the total fracture surface area due to channelling of the flow in the fracture plane (Abdalla, 2005). Other observations by Abelin et al. (1985) in the Stripa mine indicated that 5-20 % of the fracture apertures carried more than 90 % of the flowing water.

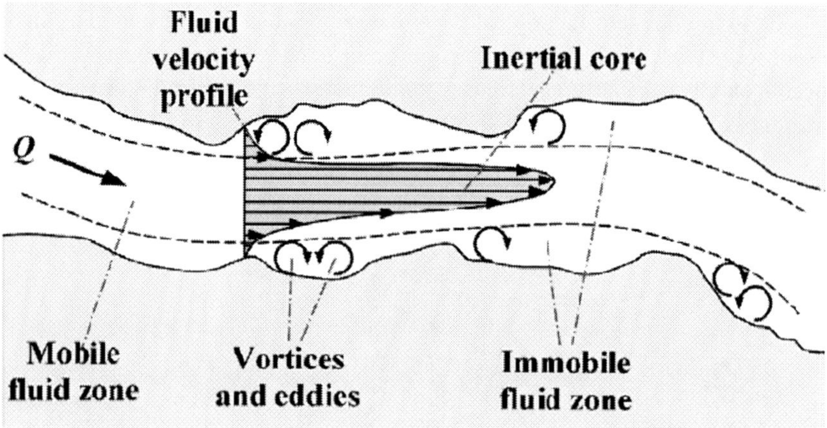

Figure 3.7. Zones of mobile and immobile water in a fracture (after Raven et al, 1988).

Therefore, small-scale laboratory experiments may significantly overestimate the effective porosity and specific surface area unless network effects are taken into account (Singhal & Gupta, 1999). This will lead to overestimation of the capacity of retardation by means of sorption and matrix diffusion (Dverstorp et al, 1992).

3.2.4. Diffusion in Fractured Media

3.2.4.1. Matrix Diffusion: General Concepts

The effects of molecular diffusion on contaminant behaviour in fractured rock aquifers were studied for the first time by Foster (1975).

Matrix diffusion is a transport mechanism by which solutes are transferred from the water-flowing channels within the fractures to the non-flowing regions (stagnant water zones in between channels as well as rock matrix), and vice versa due to the concentration gradients (Abdalla, 2005) (figure 3.1, 3.8).

The phenomenon of diffusion will be more relevant in flow paths which have a larger exposed rock surface; the inner surfaces in the rock matrix are many times (thousands or more) larger than the surfaces of the fractures, resulting in solute's accessibility, through diffusion, to a large volume of voids and sorption sites within the matrix (figure 3.9). Moreover, the presence of fracture coatings may also create a zone of diffusion having different diffusion coefficients from those in the unaltered matrix, with the effect of further delaying solute transport.

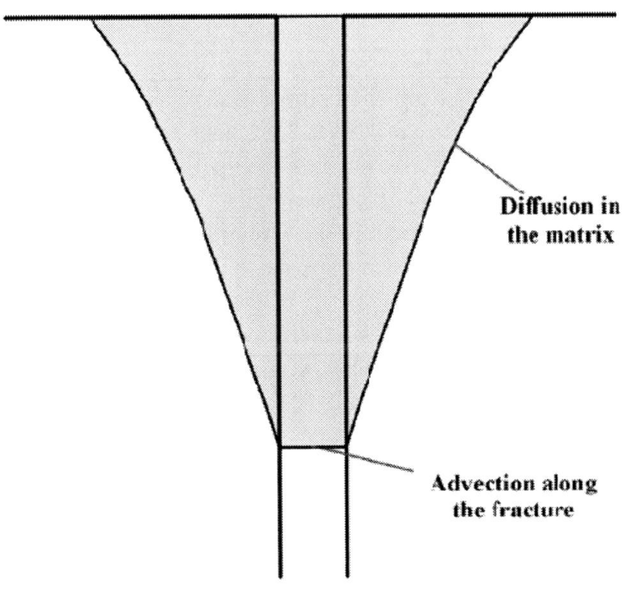

Figure 3.8. Schematization of transport in a single fracture showing solute advection along the fracture and a diffusion halo into the porous matrix.

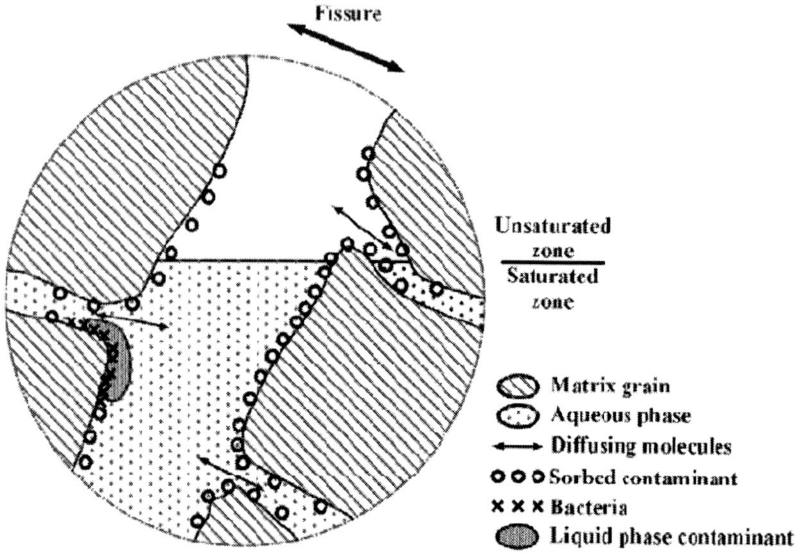

Figure 3.9. Schematic diagram of diffusion across a fissure-matrix boundary (Lyman et al, 1992).

The process of molecular diffusion is slow: in high permeability formations, characterized also by non porous matrix, such as for example tight fractured granite, its effects are masked by higher velocities; instead, in low permeability material, such as clay or silt, where the matrix is porous and groundwater velocities are low, it can have significant effects on contaminant propagation (Singhal & Gupta, 1999). A retardation effect of about three orders of magnitude in contaminant propagation is reported from fractured clays (Germain & Frind, 1989)

Diffusion of contaminants in fractured rocks has been studied for various fracture openings, spacings and other fracture geometries by a number of researchers (Sudicky & Frind, 1982; Vogel & Giesel, 1989; Rowe et al, 1989). These studies have agreed that the parameter fracture aperture is inversely proportional to diffusion: Large fracture apertures or high velocities will reduce matrix diffusion due to increase in the solute-transported to solute-stored ratio (Singhal & Gupta, 1999).

As far as the relationship diffusion/fracture spacing, there are contrasting results: for some researchers the two parameters are directly proportional; small fracture spacings will provide a greater penetration distance along the fractures because of the limited capability of the porous matrix to store solute, resulting in insignificant molecular diffusion in the matrix (Grisak & Pickens, 1980; Sudicky & Frind, 1982). Rowe et al. (1989) showed that, in fractured shales, matrix diffusion effects are negligible for fracture spacings of 0.01 or less, but they can appreciably alter the shape and extent of the contaminant plume for wider fracture spacings (e.g. 1 m).

On the contrary, for other scientists a small fracture spacing will promote rapid diffusion phenomena, allowing fractures and matrix to establish equilibrium (see par. 3.2.5).

Diffusion from the fractures into the porous matrix is a physical retardation mechanism with the effect of tailing on contaminant breakthrough curves (Abdalla, 2005, Singhal & Gupta, 1999).

Diffusion coefficients in consolidated sedimentary rocks mostly range between 10^{-11} and 10^{-14} m^2 s^{-1} (Cook, 2003, Parker et al., 1994). Figure 3.10 shows the relationship between solute diffusion coefficient and matrix porosity for core samples of Lincolnshire Limestone, that emphasizes the presence of a direct proportionality between the two parameters. Moreover, a number of studies have observed increased matrix porosities and diffusion coefficients in weathered rock adjacent to fractures (Cook, 2003). For example, the measurement of matrix diffusion coefficients and matrix porosity values for granite samples gave significantly higher values at the fracture surface than at a distance of 5 cm from the fracture (Bradbury & Green, 1985). A zone of increased porosity of 10–20 cm was measured at either side of a fracture in the Lincolnshire Limestone (Cook, 2003; Greswell et al., 1998).

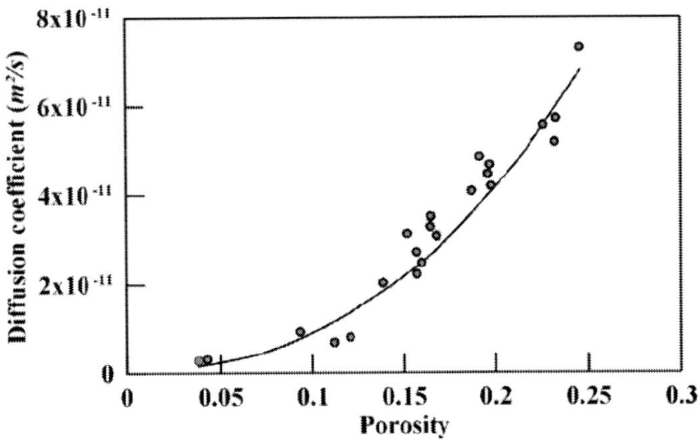

Figure 3.10. Relationship between solute diffusion coefficient for KBr and matrix porosity for core samples of Lincolnshire Limestone (Greswell et al., 1998).

Other studies have confirmed the direct proportionality between porosity and diffusivity in that they proved that in a fractured rock mass having a not negligible primary porosity, the diffusive process may be significant. The showed graph (figure 3.11) uses the solution of the solute transport equation in a fracture of 0.5 mm aperture, in order to calculate the breakthrough tracer concentration C/C_0 where C_0 is the initial tracer concentration.

The input of concentration is continuous and the point of measurement is located 40 meter away from the injection point.

The breakthrough curves show that, for high velocities (10 m/day) and low porosities(1%) the diffusive phenomenon is negligible, while it starts having influence on contaminant transport if the porosity of the matrix increases (5% to 10%).

If also velocity decreases (up to 1 m/day), for matrix porosity values of 10%, the breakthrough curve shows a strong influence of diffusive phenomena on solute transport, in that very low values of C are registered at the point of measurement.

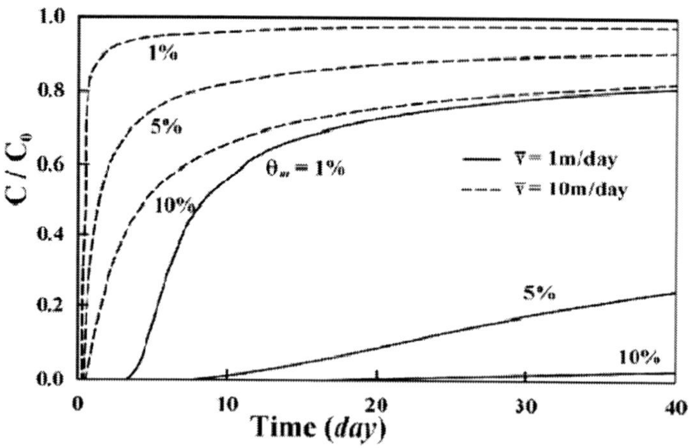

Figure 3.11. Breakthrough tracer concentration curves in a 0.5 mm fracture varying velocity and matrix porosity (Tsang & Tsang, 1987).

3.2.4.2. Equations for Diffusion

The matrix to secondary pore channel diffusion can be described by the following equation, derived from Fick's first law, which describes the mass of chemical expected to diffuse into the rock matrix from a macropore per unit area (Skagius & Neretnieks, 1986):

$$M = \frac{C_l D_e}{l} t - \frac{C_l l \alpha}{6} \tag{56}$$

where

M= mass of diffused substance, at time t, per unit diffusive surface area (mg/m^2)

l = distance of diffusion (normal to macropore surface)(m)

C_l = concentration in macropore water (mg/m^3)

t = time (s)

D_e = effective diffusion coefficient, or effective diffusivity (m^2/s)

α = rock capacity factor (dimensionsless)

The rock capacity factor α measures the rock matrix capacity to store, instead of transmit, as a function of both porosity and sorption (Lyman et al, 1992):

$$\alpha = n + \rho K_d \tag{57}$$

K_d distribution coefficient (for non-absorbable solute = 0) $[L^3M^{-1}]$

If the diffusing component is non-adsorbed α will equal the total matrix porosity (Skagius & Neretnieks 1986).

The total porosity n is made up of transport porosity or effective porosity n_t and storage porosity n_d as follows:

$$n = n_t + n_d \tag{58}$$

The transport porosity n_t describes those pores that are connected and used for transport in x-direction. The storage porosity n_d refers to pores that have a dead end. Hence the storage porosity does not contribute to the transport process itself, but it can affect the capacity to hold the contaminants; therefore it can be neglected in the transport equations.

The effective diffusion coefficient D_e that describes the local matrix diffusion phenomena is smaller than the water molecular diffusion D_m, as the matrix itself partially delays and impede the solute flux because of:

- The transport porosity n_t, describing the cross sectional area occupied by the solute

- The tortuosity factor τ (figure 3.12a), a purely geometric parameter that accounts for the increased distance of transport. It is usually expressed as the square of the ratio of the effective path length l_e to the shortest distance l (Abdalla, 2005). Hence $l_e > l$ for porous media, τ is always greater than 1 (Epstein 1989):

$$\tau = \left(\frac{l_e}{l}\right)^2 > 1 \tag{59}$$

The value of τ varies from 0.1 in clayey sands to 0.7 in sands; in crystalline rock τ is < 0.02 (UNESCO, 1980).

- The constrictivity. The cross sectional area that is accessible for diffusive transport varies along a pore, as illustrated in figure 3.12b. The most constricted parts of the pores limit the diffusive flux, resulting in that the entire pore volume cannot be effectively used for transport (Löfgren, 2004). The term δ represents the constrictivity of pores.

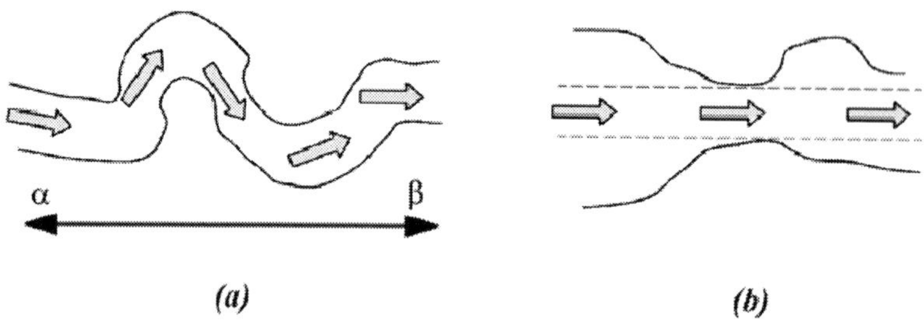

Figure 3.12. Illustration of the effect of a) tortuosity and b) constrictivity on matrix diffusion (Löfgren, 2004).

The tortuosity τ_2 (-) and constrictivity δ (-) of the porous system further reduce the transport capacity of the rock:

$$D_e = D_m n_t \frac{\delta}{\tau^2} \tag{60}$$

D_m diffusion coefficient in free solution $[L^2T^{-1}]$

D_e effective diffusion coefficient $[L^2T^{-1}]$
δ/τ^2 geometric factor [-]

The effective diffusion coefficient measures a rock' s capacity to transmit by diffusion, and is a combination of diffusivity in the primary porosity and the portion of the porosity that is devoted to transport (Lyman et al, 1992):

$$D_e = D_p n_t \tag{61}$$

D_p pore diffusion coefficient $[L^2T^{-1}]$

At small scale (few connected pores), the pore diffusion coefficient D_p includes the resistance to diffusion from the pore geometry δ/τ^2.

At large scale (rock matrix), the effective diffusion coefficient D_e includes the effects of porosity through a formation factor F [-]. If the diffusing species do not interact with the rock matrix and are small compared to the pore aperture, the formation factor F is a geometrical factor that is only dependent on the geometry of the micropore network and is independent of the diffusing species (Löfgren, 2004):

$$F = n_t \frac{\delta}{\tau^2} \tag{62}$$

The effective diffusion coefficient can be expressed as:

$$D_e = D_p n_t = D_m n_t \frac{\delta}{\tau^2} \tag{63}$$

$$D' = \frac{D_e}{D_m} = n_t \frac{\delta}{\tau^2} = F \tag{64}$$

D' relative diffusivity [-]

According to Archie's law the relative diffusivity D' depends only on the diffusive properties of the rock samples and is considered to be independent of temperature, sorption capacities of the porous media and the properties of the solute (Abdalla, 2005). Archie's law permits to avoid the realization of time-consuming diffusion experiments for the prediction of diffusion coefficients and to use instead a relation valid for any kind of tracer or contaminant.

To sum up, it is possible to establish a relation of inequality between the different types of diffusion coefficients:

$D_m > D_p > D_e > D_a$ (65)

D_a apparent diffusion coefficient $[L^2T^{-1}]$

The change in concentration of an adsorbing solute in the porous media under transient conditions is described by Fick's 2nd law:

$$(n + \rho K_d)\frac{\partial C_m}{\partial t} = D_p n_t \frac{\partial^2 C_m}{\partial z^2} \tag{66}$$

$$\frac{\partial C_m}{\partial t} = \frac{D_p n_t}{(n + \rho K_d)} \frac{\partial^2 C_m}{\partial z^2} \tag{67}$$

ρ density $[ML^{-3}]$

C_m concentration in the porous matrix $[ML^{-3}]$

$\partial C_m / \partial x$ concentration gradient (negative in the direction of diffusion) $[ML^{-3}T^{-1}]$

It is possible to synthesize the term $D_p \frac{n_t}{(n + \rho K_d)}$ into a unique term, D_a that is called the generic (apparent) diffusion coefficient of Fick' s second law, and is equal to $D_a = D_e/\alpha$ with D_e and α being independent of one another (Lyman et al, 1992). No correlation between these two parameters has been demonstrated in the experiments carried out by Skagius and Neretnieks (1986). So the previous equation becomes:

$$\frac{\partial C_m}{\partial t} = D_a \frac{\partial^2 C_m}{\partial z^2} \tag{68}$$

Where:

$$D_a = D_p \frac{n_t}{(n + \rho K_d)} \tag{69}$$

If the species interact strongly with the rock matrix, the diffusive flux through the rock becomes mainly dependent on sorption processes and not on the restrictions due to the geometry of the porous system (Löfgren, 2004). The formation factors of intrusive igneous and metamorphosed intrusive igneous rock from a number of sites in Sweden are reported to range from 10^{-7} to 10^{-3} (Skagius, 1986).

In order to account for diffusivity, Cook (2003) made use of an approximate solution of the equation of Lever & Bradbury (1985) for the transport of a short pulse of solute through a single fracture; according with this solution, the distance reached by a tracer, released into a fracture, after a given period of time, t can be expressed:

$$x = V_f \frac{b}{2}\sqrt{\frac{t}{D_e n_m}} \tag{70}$$

where V_f is the water velocity within the fractures and D_e is the effective diffusion coefficient within the matrix, b the fracture aperture. Given a value of $V_f = 35$ m/day, b= 250 μm, $n_m = 0.05$ and $D_e = 10^{-4}$ m^2/year, t= 1 year, the x value will be equal to 1386 m. By contrast, the distance traveled by groundwater will be approximately 13 km, almost one order of magnitude higher, due to the absence of diffusive phenomena. Moreover, it should be pointed out that distance travelled through a fracture is proportional to the square root of time, differently from porous media where the distance travelled by a solute is directly proportional to the travel time. This implies that in a fracture if the solute travels 10 m in the first year, it will travel a further 4 m (and not 10 m) in the following year (Cook, 2003).

3.2.5. Transport by Advection and Diffusion in Fractured Aquifers

Cook (2003) illustrated the relationship between water velocities and apparent solute velocities by considering two end-member scenarios. The first considered scenario is without diffusion in the matrix, that implies that the tracer velocity is equal to the water velocity through the fractures (figure 3.13a). The opposite case takes place when diffusion is very rapid and fractures are spaced very closely together, so the solute concentration in the matrix equilibrates with the one within fractures (figure 3.13d).

This scenario occurs generally when D_e/B^2 is large (that means fracture spacing is small relative to the matrix diffusion coefficient) and is referred to as *equivalent porous media* (*EPM*) condition. In this case, the apparent tracer velocity V_{EPM} will be equal to the water velocity in the fractures, V_f, divided by the ratio of the total porosity, n_t, to the fracture porosity, n_f (Cook, 2003):

$$V_{EPM} = V_f \frac{n_f}{n_t} \tag{71}$$

In-between these two scenarios (figure 3.13c), the apparent solute velocities, V_a, will be less than the water velocity in the fractures, but greater than V_{EPM}.

$$V_{EPM} < V_a < V_w \tag{72}$$

If on the contrary the ratio D_e/B^2 is small (that means matrix diffusion coefficient is small relative to fracture spacing), then diffusion into the matrix from one fracture penetrates less than half-way to the adjacent fractures (figure 3.13b) (Cook, 2003).

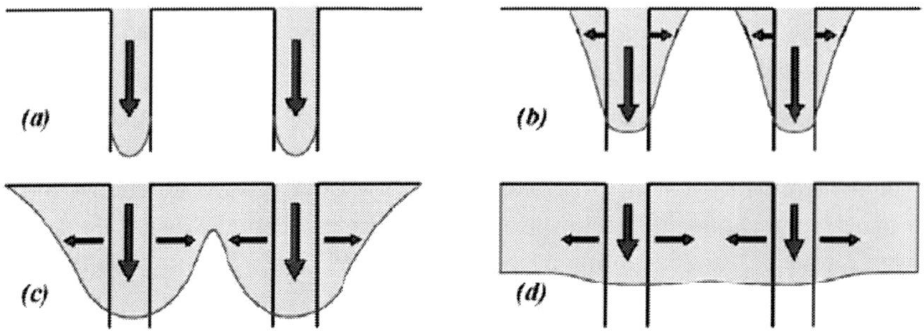

Figure 3.13. Tracer movement through fractured rocks with (a) no diffusion into the rock matrix; (b) diffusion penetrating less than half-way between adjacent fractures; (c) diffusion penetrating more than half-way between adjacent fractures; (d) Complete equilibration of fracture and matrix concentrations (Cook, 2003).

3.3. Reactive Solute Transport

3.3.1. Overview

In addition to the physical transport mechanisms discussed in the previous section, there are a wide range of reactive transport mechanisms that occur within a fracture, along the fracture walls, and/or within the adjacent porous rock matrix. These processes include adsorption/desorption, dissolution/precipitation, radionuclide decay, organic reactions and volatilization, and/or biodegradation (Berkowitz, 2002). Rocks may have secondary clay mineral coatings, located in fractures, or may contain primary minerals which could catalyze chemical oxidation. Ion exchange or adsorption reactions are of importance in the movement of contaminants in groundwater systems because they play an important role in the slowdown and retardation of solutes in the aquifers (figure 3.1).

3.3.2. Sorption

Sorption reactions are very rapid relative to the flow of groundwater; they are generally considered as instantaneous and reversible processes.

The entity of sorption processes depends a) on groundwater flow velocities b) on dimension of surface area of the medium c) on physical and chemical composition of surface area of the medium.

As far as the first aspect is concerned, in case of high groundwater velocities sorption mainly occurs at the surfaces of the fractures. With increasing residence times due to smaller velocities, sorption occurs within the matrix due to the available exchange surfaces in the matrix (Abdalla, 2005).

However, in general, the rate of sorption will be much greater by the matrix material due to its greater surface area as compared with the fracture surface.

The extent of sorption depends also on the nature of matrix material. In cases the matrix has negligible porosity, as in a purely fractured rock, sorption will take place only on the surface area of the fractures which is relatively small. However, in fractured porous rock, adsorption will also take place on the very large solid surface area within the porous blocks (Singhal & Gupta, 1999). In the latter case, the entity of adsorption is also affected by type of ions and composition of matrix material: it may contain colloidal-sized particles (< 2 μm) such as clays, organic matter, iron and aluminium oxides and hydroxides that sorb pollutants from groundwater.

Freeze and Cherry (1979) developed a simple model accounting for adsorption along planar fracture walls. For fractured media, the distribution coefficient K_d can be used to compute the retardation coefficient of fracture surface R_f and retardation coefficient of rock matrix R_m. According to Henry sorption isotherm, the retardation equation can be expressed as follows (Freeze & Cherry, 1979):

$$R_f = 1 + a_w K_{d_f} = 1 + \frac{K_{d_f}}{b} \tag{73}$$

$$R_m = 1 + \frac{\rho K_d}{n_m} = \frac{\alpha}{n_m} \tag{74}$$

where:

K_d matrix distribution coefficient $[L^3M^{-1}]$
K_{df} distribution coefficient for the fracture surface [L]
a_w specific fracture surface $[L^{-1}]$
ρ_b dry bulk density of the dry matrix material $[ML^{-3}]$ densità di nucleo della matrice rocciosa
n_m matrix porosity

The equation is valid if the fracture aperture b can be determined and the fracture surface is assumed to be planar, such as to have $a_w = 1/b$.

The distribution coefficient for the fracture surface K_{df} is defined as the mass of solute adsorbed per unit area, divided on solute concentration in water. In order to put in relation this coefficient with its homologous for rock matrix, K_d that is based on bulk density of matrix material, it is necessary to estimate preliminarily the inner specific surface of the medium, γ, defined as the area of the pore walls exposed to fluid, having the measurement unit of area per unit mass. In this way, the relation between K_{df} and K_d can be expressed:

$$K_d = \gamma K_{df} \tag{75}$$

For non-reactive solutes such as chloride K_d equals zero and R_m one (Maloszewski & Zuber, 1990; Shackelford, 1991; Kozaki et al., 1999; Maloszewski et al., 1999).

Equation [73] indicates that retardation decreases with increase in fracture aperture (Singhal & Gupta, 1999).

Although the retardation equation is simple in conceptual terms, it is difficult to apply to field systems because it assumes an uniform value of the retardation coefficient. This assumption is unrealistic in case of heterogeneous fractured media, characterized by variation in fracture aperture and orientation; fracture surfaces usually are non-planar, and without elaborate experimental effort, the actual surface area with which the contaminant reacts is unknown (Freeze & Cherry, 1979). More sophisticated models must be developed to account for sorptive transport in fractured media in a computationally tractable manner. In general, the state-of-the-art is to assume a single, overall retardation coefficient for the fractured system (Berkowitz, 2002).

According to Lyman et al (2000), there is a relationship between the degree of sorption and diffusivity. In the definition of the generic (or apparent) diffusion coefficient D_a is included both the capacity to transmit, reflected by D_e, and the capacity to store, reflected by α. Even if not much documented in the literature, it is conceivable that a relationship between the two variables exists if sorption takes place in such a manner as to remove the sorbed molecules from the concentration gradient which is "felt" by subsequently diffusing molecules (Lymann et al, 2000).

3.3.3. Effects of Advection, Diffusion and Adsorption

In figure 3.14 are shown the effects of advection, diffusion and adsorption on solute transport in fractured media, characterized by a non-porous and a porous matrix. Profiles of relative concentration (C/C_o) of the contaminant along the fracture at time, t_1 in these cases, are also shown. In figure 3.14a, the matrix is non-porous and dispersion in the fracture is neglected, therefore the concentration profile has a rectangular shape. On the contrary in figure 3.14b, where the matrix is porous, molecular diffusion is prominent and retards the advance of solute along the fracture. In figure 3.14c the added effect of adsorption in the surface of the fracture causes further retardation in the migration of contaminant in the fracture (Singhal & Gupta, 1999).

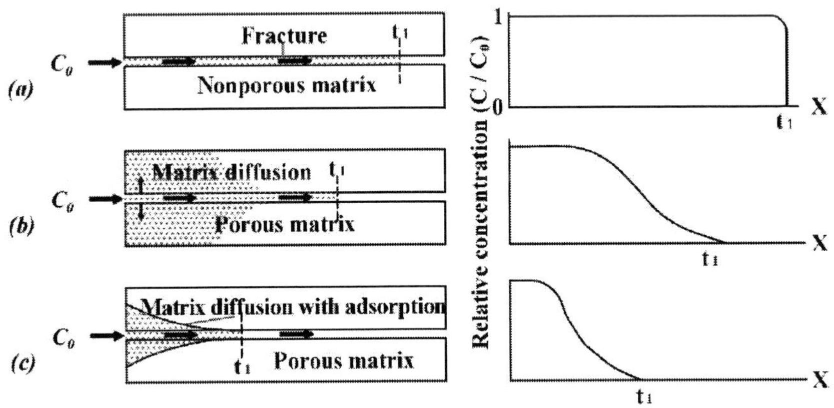

Figure 3.14. (a) Advective transport along a fracture in a nonporous matrix; (b) advective transport along a fracture with diffusion in the porous matrix; (c) advective transport along a fracture with diffusion and adsorption in the porous matrix (Singhal & Gupta, 1999).

3.4. Influence of Fracture Rock Parameters on Remediation

Figure 3.15 shows the decrease in apparent solute velocity with time calculated for V_w = 500 m/year, n_m = 0.1, B = 0.05 m, b = 50 μm, D_e = 10^{-5} m^2/yr. Initially, the velocity decreases with the square root of time (Eq. 70) as solute diffuses into the rock matrix. At very late times, diffusion may result in equivalent tracer concentrations within the fractures and the aquifer matrix (Equivalent Porous Media condition) (Cook, 2003).

When this occurs, the velocity approaches a constant value, and then the distance travelled by a solute will increase directly with time as is the case for porous media aquifers. The velocity trend shown in figure 3.15 is partially reflected by a typical behavior of contaminant breakdown in a fracture in time after the application of a remediation technique (figure 3.16): initially a fast drop in concentration is shown followed by a long period in which the concentration values decrease more slowly.

This behavior is proper of a fractured rock aquifer in which matrix diffusion phenomenon strongly influences solute transport mechanisms. However, several rock mass properties exert a more or less relevant influence on cleanup rate.

In this optic, a study of Mutch & Scott (1993) reported here, analyses the effects of different aquifer parameters on groundwater clean-up for fractured media. They carried out a sensitivity analysis based on determining aquifer clean up time varying each single parameter (over a range of values taken from literature) while keeping the others fixed in a numerical model used to simulate remediation of a fractured aquifer.

The results of this study showed that not all parameters exert the same influence on contaminant breakdown. According to Mutch & Scott (1993), aquifer parameters can be divided into three groups: the first group is characterized by parameters that have a very relevant effect on the cleanup rate, such as matrix diffusivity and fracture spacing; the second group, formed by parameters that have a relevant influence on the initial drop in concentration but not on the ultimate cleanup rate, such as matrix porosity, groundwater flow rate, fracture length, length of the period of contamination; finally the third group is made up of parameters, such as fracture porosity and longitudinal dispersivity, that have essentially no relevant effect on the cleanup rate.

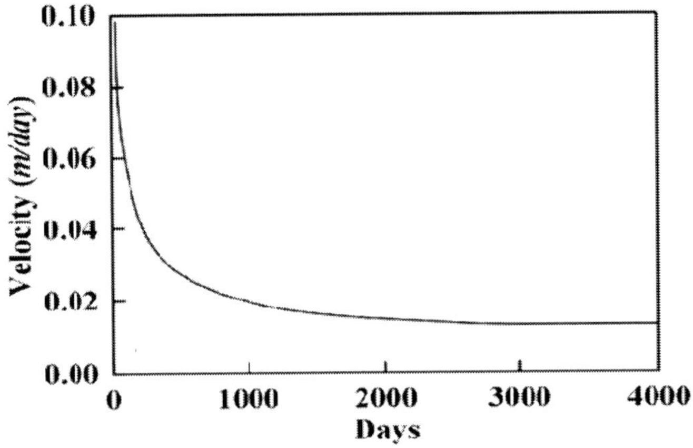

Figure 3.15. Apparent velocity of a tracer in fractured rock (Cook, 2003).

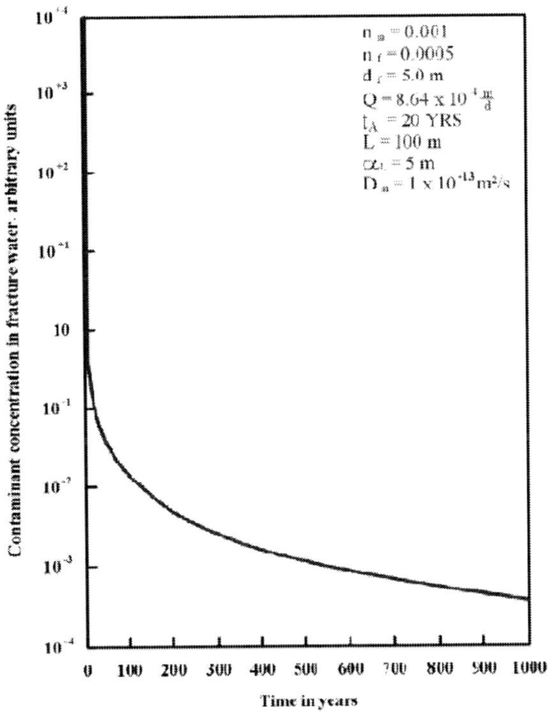

Figure 3.16. Contaminant breakdown in the fracture after remediation (Mutch & Scott, 1993).

The diffusivity affects strongly both the initial rate of decrease and the final linear rate (figure 3.17); a slower rate of initial cleanup associated to a much faster ultimate cleanup is typical of higher diffusivities; this is due to the contaminant' s diffusion back into the fracture at a much higher rate, that implies not only higher initial concentration in the fracture but also faster ultimate removal from the matrix. Analogous effects have been registered as far as fracture spacing; for example increasing fracture spacing leads to similar results to reducing diffusivity: a higher initial and a slower final cleanup rate. On the contrary, a small fracture spacing can be associated with a very high diffusivity, as it shows a very rapid final cleanup.

This analysis has shown that both matrix diffusivity and fracture spacing are crucial parameters within the cleanup procedure in an aquifer; numerical models used to simulate remediation processes make use of averaged values for these two parameters, as it is extremely difficult to determine them accurately. Therefore, making predictions of actual cleanup times in fractured aquifers is most of the times of scarce significance.

As far as the second group of parameters, in figure 3.18 are compared the effects on cleanup rate of varying matrix porosity and length of the contamination period. These parameters affect only the initial drop in concentration; low matrix porosity implies a faster initial cleanup rate, because there is less immobile water for contaminant to diffuse into or out of the matrix itself, and therefore a high percentage of solute remains in the fracture water and can be easily removed.

Despite the initial flushing period, the curves then become nearly parallel, that means that each order-of magnitude variation in concentration needs about the same period of time for each value of matrix porosity. Analogous effects have been obtained for the length of contamination period (figure 3.18b), groundwater flow rate and fracture length (not shown).

Finally, considering this analysis, it is possible to explain the trend of the cleanup rate in fractured rock aquifers (figure 3.16): as they normally have low matrix diffusivities and porosities and wide fracture spacings, it is straightforward that the curve that represents the trend of breakdown of contamination is characterized by a very fast initial and extremely slow final cleanup rate.

Therefore, in a context where remediation rates are affected by the process of matrix diffusion, groundwater extraction and treatment systems are not fully effective in speeding up the aquifer's remediation. Pump & Treat system has proven to be efficient in the contaminant breakdown as long as it is in operation, but after the cessation of pumping a rebound effect brings pollutant concentration values rapidly to nearly no-action levels (Mutch & Scott, 1993).

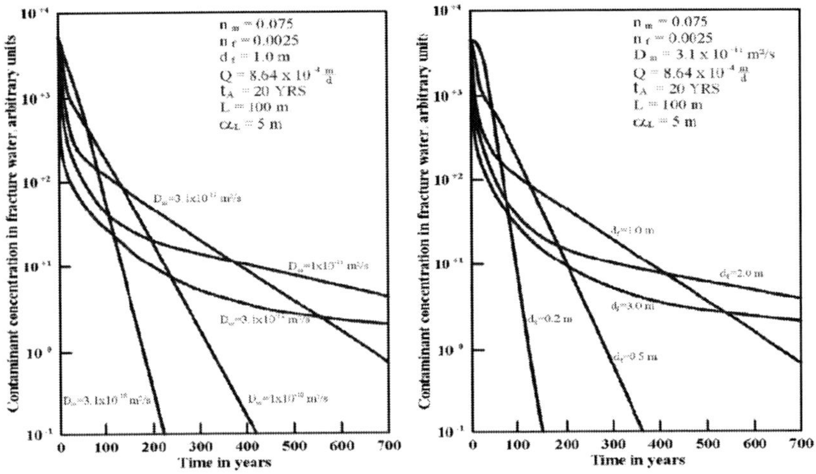

Figure 3.17. a) effect of matrix diffusivity and b) fracture spacing on remediation rate in fractured aquifers (Mutch & Scott, 1993).

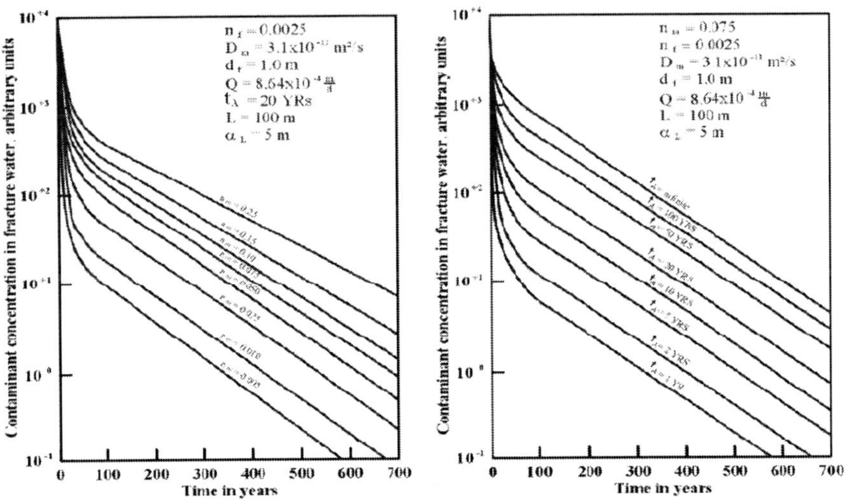

Figure 3.18. a) effect of matrix porosity and b) length of the contamination period on remediation rate in fractured aquifers (Mutch & Scott, 1993).

Even if not directly investigated by the authors, the parameter matrix hydraulic conductivity that normally has negligible values can significantly influence cleanup rates in the cases in which it is high enough to create even a low flow rate through the rock matrix (Mutch & Scott, 1993). Therefore, in absence of a natural advective component of flow in the matrix, cleanup rates may be increased significantly by inducing such flow.

The extraction system has to be carefully designed in such a way as to create a relevant pressure gradient across the porous matrix and therefore speed up markedly transport of contaminant out of it. Another possible operative way is to accelerate the diffusion rates out of the rock, by means of reducing fracture spacing (figure 3.17b).

This can be achieved, during pumping, by forcing flow through small fractures which are not part of the natural fracture flow system. The opening and flushing of additional fractures is also achieved by techniques such as hydrofracturing, pneumatic fracturing or explosive blasting that are successful in increasing the degree of fracturing and containing the plume, provided that such increased fracturing is achieved over the entire volume of the plume, not just downgradient to it.

Another way to accelerate cleanup processes in a diffusion controlled system is dewatering the rock matrix, on order to make contaminants out of dry rock by the process of gaseous diffusion. This technique requires first of all dewatering the fracture flow system, but also maintaining a low moisture level in the fractures in order to make water evaporate from the rock matrix.

When dealing with DNAPLs, a commonly used remediation technique in fractured aquifers consists in dewatering the contaminated zone via high-rate pumping and then applying Soil Vapour Extraction to remove volatile DNAPLs, or applying steam to mobilize and volatilize the DNAPL towards a collection well (Keller, 1999).

REFERENCES

Abelin, H., Birgersson, L., Gidlund, J. and Neretnieks, I. (1991) A Large-Scale flow and tracer experiment in Granite. 1. Experimental design and flow distribution. *Water Resour. Res.* 27 (12): 31o7-3117.

Abelin, H., Neretnieks, I., Tunbrant, S., and Moreno, L. (1985) Final report of the migration in a single fracture- *Experimental results and evaluation, Tech. Rep.* 85-03, Stripa Proj., Stockholm.

Abdalla, F.A. (2004) Contaminant Transport In *A Fractured Chalk Aquifer At Sigerslev, Denmark,* As Characterised By Tracer Techniques Dissertation zur Erlangung des Doktorgrades (Dr. rer.nat.) der Mathematisch-Naturwissenschaftlichen Fakultät der Rheinischen Friedrich-Wilhelms-Universität Bonn vorgelegt von FATHY AHMED ABDALLA AHMED aus Qena Ägypten Bonn, Dezember 2004.

Atkinson, B. K., (1987) 'Introduction to fracture mechanics and its geophysical applications', *Fracture Mechanics of Rock,* (ed. B. K. Atkinson), Academic Press, London, pp. 1-26.

Bai M. & Roegiers J.C., (1995) Modeling of heat flow and solute transport in fractured rock masses. *Proc. 8th Int. Congress on Rock Mech.,* Japan.

Bai M. & Roegiers J. C. (1997) 'Triple-porosity analysis of solute transport', *Journal of Contami nant Hydrology.* 28, 247–266.

M. Bai, Roegiers J. C. and Elsworth D. (1995) Poromechanical response of fractured-porous rock masses, *J. Pet. Sci. Eng.* 13, 155-168.

Bai M., Roegiers, J.C. and Inyang, H.I. (1996) Contaminant transport in non-isothermal fractured media. *Journal of Environmental Engineering, ASCE,* 122(5), 416-423.

Bai M., Inyang H.I., Chien C.C. and Bruell C. (2000) Factors for assessing flow and transport in fractured porous media. In: *Remediation in Rock Masses,* American Society of Civil Engineers, New York, published by Inyang H. I. & Bruell C. J. (2000) ISBN 0-7844-0015-6 (2000), 250 pages.

Barton, N.R. (1973) Review of a new shear strength criterion for rock joints. *Engng. Geol.* 7, 287-332.

Barton N. & Choubey V. (1977) The shear strength of rock joints in theory and practice. *Rock Mechanics.* 10:1–54.

Batchelor G.K. *(1967)* An Introduction to Fluid Dynamics, *Cambridge University Press.* ISBN 0521663962.

Bear, J. (1972) *Dynamics of Fluids in Porous Media.* 764 pp. New York, London (Elsevier).

Bear J., Tsang C.F., de Marsily G. (1993) *Flow and contaminant transport in fractured rock.* Academic Press, San Diego, CA.

Becker, M. W. & Shapiro, A. M. (2000) Tracer transport in fractured crystalline rock: Evidence of nondiffusive breakthrough tailing. *Water Resour. Res.* 36 (7): 1677-1686.

Beggs K.A., Gustason, E.R. and Angeit P.F. (1993) Characterizing and managing the dynamic reservoir - a multi disciplinary approach: *Characterization of a complex fluvial - deltaic reservoir for simulation.* 4 h Annual Archie Conference. Texas, p. 143-148.

Berkowitz, B. (2002) Characterizing flow and transport in fractured geological media: *A review. Advances in Water Resources*. Volume 25, Issues 8-12, August-December 2002, Pages 861-884.

Black, J.H. (1987) Flow and flow mechanisms in crystalline rock, Fluid Flow in Sedimentary Basins and Aquifers (eds J.C. Goff and B.P.J. Williams) *Geol. Soc. Spl. Publ.* No. 34, pp 185-200.

Bordas J.M. (2005) Modeling Groundwater Flow And Contaminant Transport In Fractured Aquifers *Thesis Faculty Department of Systems and Engineering Management Graduate School of Engineering and Management Air Force Institute of Technology Air University.*

Borgia G.C., Bortolotti V. and Masciopinto C. (2002) Estimation of contribution of effective fractured rock porosity to the aquifer transmissivity by means of both lab methods and field tests (in Italian with abstract in English and French: *Valutazione del contributo della porosità effettiva alla trasmissività di acquiferi fratturati con tecniche di laboratorio e di campo). IGEA, Groundwater Geo-engineering,* 17, 31-43.

Bouhroum, A. & Civan, F., (1994) Study of particulates migration in gravel pack," Proceedings, SPE Int. *Symp. for Damage Control, Society of Petroleum Engineers,* pp. 75-91, 1994.

Bradbury, K R., M. A. Muldoon, A. Zaporozec, and J. Levy, (1991) Delineation of wellhead protection areas in fractured rocks. US EPA, *Office of Ground Water and Drinking Water*, 144p.

Brown M. (2005) Characterization of a Diesel-Contaminated Fractured Rock Aquifer and Development of a Nutrient-Flushing Remediation Technique *Oral Presentation Remediation Technologies Symposium*. 2005.

Carlsson, A. & Olsson, T., (1977) Hydraulic properties of Swedish crystalline rocks. Hydraulic conductivity and its relation to depth. *Bulletin of the Geological Institutions of the University of Uppsala,* N.S. Vol.7, pp 71 – 84. Uppsala.

Caswell B. (1992) Protecting fractured-bedrock wells. *Water Well Journal.* 46(5):42–45.

Cherubini C. Ghezzehei T. A and Su G. W. (2008) The drift shadow phenomenon in an unsaturated fractured environment. In: D.G. Toll, C.E. Augarde, D. Gallipoli, S.J. Wheeler. *Unsaturated Soils.* Advances in Geo-Engineering. ISBN: 978-0-415-47692-8. FLORENCE KENTUCKY: Taylor & Francis Books (UNITED STATES).

Chirlin, G.R. (1985) Flow through a porous medium with periodic barriers or fractures. *Society of Petroleum Engineers journal.* vol. 25, n°3, pp. 358-362.

Clauser C. (1992) *EOS, Transactions American Geophysical Union*, Vol. 73, NO. 21, PAGE 233, 1992 Permeability of crystalline rocks.

Clerici A. & Sfratato F., (2008) Stima della conducibilità idraulica in ammassi rocciosi Italian *Journal of Engineering Geology and Environment, Special Issue 1.* (2008).

Cook P.G., Love A.J. and Dighton J.C. (1999) Inferring ground water flow in fractured rock from dissolved radon. *Ground Water.* 37(4):606–610.

Cook, P.G. (2003) A Guide To Regional Groundwater Flow In *Fractured Rock Aquifers,* contributions from Williams, M., Simmons, C.T., Love, A.J., Halihan, T., Herbert, G., and Heinson, G., 2003., Seaview Press, Henley Beach, South Australia, 108 pp. ISBN: 978-1740082334.

De Marsily, G. (1986) *Quantitative Hydrogeology,* Academic Press, New York, p.440.

Dershowitz, W.S. & H.H. Herda (1992) Interpretation of Fracture Spacing and Intensity. *Proc. 32nd US Rock Mech. Symp., Santa Fe, NM,* p.757-766.

Dverstorp B., J. Andersson J. and Nordqvist W. (1992) Discrete fracture network interpretation of field tracer migration in sparsely fractured rock, *Water Resour. Res.* 28(9) (1992) 2327-2343.

Einsiedl, F., Maloszewski, P., Witthüser, K. and Wohnlich, S. (2000) Application of two new fluorescent dyes and fluorescent particles in a horizontal and vertical fracture. In: *Dassargues, A. (Ed.) (2000): Tracers and Modelling in Hydrogeology.- IAHS Publication* No. 262: 175-179; Wallingford, UK (IAHS Press).

Freeze, R.A. and Cherry, J.A. (1986) *Groundwater,* Prentice Hall, New Jersey.

Fetter, C.W. (1993) *Contaminant Hydrogeology,* Macmillan, new York, 458 pp.

Gardner, W.R. (1958) Some steady-state solutions of the moisture flow equations with application to evaporation from a water table. *Soil Science.* 85, 228-232.

Gelhar, L.W., Welty, C. and Rehfeldt, R. (1992) A critical review of data on field-scale dispersion in aquifers. *Water Resour. Res.* 28 (7): 1955–1974.

Greswell R., Yoshida K., Tellam J.H. and Lloyd J.W. (1998) The micro-scale hydrogeological properties of the Lincolnshire Limestone UK. *Quarterly J. Eng.Geol.,* 31:181–197.

Grisak, G.E. & Pickens, J.F. (1980) Solute transport through fractured media. 1. The effect of matrix diffusion. *Water Resour. Res.* 16 (4): 719-730.

Haggerty, R. (2000) Development of late-time slopes on log-log breakthrough curves after a pulse-type injection of the case of infinite matrix block. In: *Interpretations of tracer tests performed in the Culebra dolomite at the waste isolation Pilot Plant site.* Lucy, C.M., Richard, L.B. and Toya, L.J.-pp. 305-307, G.P. Sandia National Laboratories.

Healy R.W. & Cook P.G. (2002) Using groundwater levels to estimate recharge. *ydrogeology J.,* 10(1):91–109.

Ho C. K. (1997) Models of Fracture-Matrix Interactions During Multiphase Heat and Mass Flow in Unsaturated Fractured Porous Media. *Paper presented at 6th Symposium on Multiphase Transport in Porous Media, ASME International Mechanical Engineering Congress and Exposition, American Society of Mechanical Engineering,* Dallas, Texas.

Hoek E. & Bray J.W. (1981) Rock slope engineering. 3rd edition. *Institution of Mining and Metallurgy,* London.

Hsieh P. (1998) Scale effects in Fluid Flow through Fractured Geologic Media In: *Scale dependence and scale invariance in hydrology Di Garrison Sposito Cambridge University Press,* 1998 ISBN 0521571251, 9780521571258 423 pp.

Hudson J.A. & Harrison J.P. (1997) *Engineering Rock Mechanics: an Introduction to the Principles,* Oxford, Elsevier Science,, ISBN: 0-0804-1912-7.

Indraratna B. & Ranjith P. (2001) *Hydromechanical Aspects and Unsaturated Flow in Jointed Rock.* A.A. Balkema, Lisse, 286 pp.

Isherwood D. (1979) *Geosciences Data Base handbook for Modelling Nuclear Waste repository.* Vol. 1 NUREG/CR- 0912 VL, UCRL-52719, VI.

Krešić, N., (1991) Kvantitativna hidrogeologija karsta sa elementima zastite podzemnih voda [Quantitative Karst hydrogeology with elements of groundwater protection]: *Naucna knjiga,* Belgrade (Monograph; in Serbian, English abs.), 192 p.

La Pointe P.R. & Hudson J.A. (1985) Characterization and interpretation of rock mass joint patterns. *Geological Society of America Special Paper* 199. Geological Society of America, 37pp.

Lee C.H. & Farmer I. (1993) *Fluid Flow in Discontinuous Rocks.* Chapman &Hall, London p. 169.

Lever, D.A., Bradbury, M.H. and Hemingway, S.J. (1985) The effect of dead-end porosity on rock-matrix diffusion. *J. Hydrol.* 80: 45-76.

Liu, H.H., & Bodvarsson G.S. (2001) "Constitutive relations for unsaturated flow in a fracture network." *Journal of Hydrology,* 252, 116-125.

Löfgren M., (2004) Diffusive properties of granitic rock as measured by in-situ electrical methods Doctoral Thesis Department of Chemical Engineering and Technology Divisions of Chemical Engineering Royal Institute of Technology Stockholm, Sweden 2004 ISBN 91-7283-935-X.

Lomize GM. (1951) Filtratsia v treshchinovatykh porodakh (flow in fractured rocks). *Gosudarstvennoe Energetitcheskoe Izdatel'stvo,* Moskva-Leningrad,127p.

Louis C. (1968) Étude des écoulements d'eau dans les roches fissurées et de leurs influences sur la stabilité des massifs rocheurs. *Bulletin de la Direction des Études et Réchèrches. Série A.* (3): 5-132. (These presentée a l'Université de Karlsruhe).

Louis C. (1976) *Introduction à l'hydraulique des roches*. PhD Thesis, Paris.

Louis C. (1984) An introduction to rock hydraulics, lecture notes (non pubblicato). *SIMESCOL.* Paris, p. 16.

Louis & Maini Y. N. (1970) *Determination of in –situ hydraulic parameters in jointed rock.* Proc 2nd congress ISRM pp 1-32.

Love A.J., Cook P.G., Harrington G.A. and Simmons C.T. (2002) *Groundwater flow in the Clare Valley.* Department for Water Resources, South Australia. Report DWR02.03.0002, 43pp.

Margolin, G., Berkowitz, B. and Scher, H. (1998) Structure, flow, and generalized conductivity scaling in fractured networks. *Water Resour. Res.* 34: 2103-2121.

Montazer, P. & Wilson W.E. (1984) Conceptual hydrologic model of low in the unsaturated zone, Yucca Mountain, Nevada, *U.S. Geol. Surv. Water Resour. Invest. Rep.* 84-4355.

Moreno, L. & Tsang, C.F. (1994) Flow channelling in strongly heterogeneous porous media: A numerical study. *Water Resour. Res.* 30: 1421-1430.

Moreno L., Tsang Y. W., Tsang C. F., Hale F. V., Neretnieks I. (1988) Flow and tracer transport in a single fracture: a stochastic model and its relation to some field observations; *Water Resour Res,* vol. 24, n° 12, pag. 2033-2048.

Mutch R.D. & Scott J. I. (1993) Problems With The Remediation Of Diffusion-Limited Fractured-Rock Systems. In: *Hazardous Waste Site Soil Remediation: Theory and Application of Innovative Technologies.* David J. Wilson, Vanderbilt University, Nashville, Tennessee, USA.

National Research Council (1996) *Rock Fractures and Fluid Flow* - Contemporary Understanding and Applications, National Academy Press, Washington DC, 551 pp.

National Research Council (2001) Conceptual Model of Vadose-Zone Transport in Fractured Weathered Shales. (In) *Conceptual Models of Flow and Transport in the Fractured Vadose Zone.* U.S. National Committee for Rock Mechanics. National Academy Press, Washington D.C. p. 87-114.

Nelson R.A. (1985) *Geological Analysis of Naturally Fractured Reservoirs: Contributions in Petroleum Geology & Engineering.* Gulf Publishing Co., Houston, TX, 320 p.

Nelson R.A. (2001) *Geological Analysis of Naturally Fractured Reservoirs.* Gulf Professional Publishing.

Nielsen K.A. (2008) *Fractured Aquifers - Formation Evaluation by Well Testing Trafford Publishing,* 2008 ISBN 1425130194, 9781425130190.

Nitao, J. J. & Buscheck T. A. (1991) Infiltration of a Liquid Front in an Unsaturated, Fractured Porous- Medium. *Water Resources Research.* 27(8) pp. 2099-2112.

Odling N.E., Gillespie P., Bourgine B., Castaing C., Chil´es J.P., Christensen N.P., Fillion E., Gender A., Olsen C., Thrane L., Trice R., Aarseth E., Walsh J.J. and Watterson J. (1999) Variations in fracture system geometry and their implications for fluid flow in fractured hydrocarbon reservoirs. *Petrol. Geosci.* 5:373–384.

Parker B.L., Gillham R.W. and Cherry J.A. (1994) Diffusive disappearance of immiscible-phase organic liquid in fractured geologic media. *Ground Water,* 32(5):805–820.

Parsons R.W. (1966) *Permeability of Idealized Fractured Rock. Soc. of Petroleum Engineers Journal,* pp. 126-136.

Pawley S. (2002) *Influence of Fracture Geometry on Pumping Well Capture Final Year Environmental Engineering Thesis University of Western Australia.*

Peters, R. R., & Klavetter E.A. (1988) A continuum model for water movement in an unsaturated fractured rock mass. *Water Resour. Res.* 24: 416-430.

Prats M. (1972) The influence of oriented arrays of thin impermeable shale lenses or of highly conductive natural fractures on apparent permeability anisotropy. *J. Pet. Technol.,* 24(10):1219–1221.

Price J. R. (2005) *Coupled analysis of two phase flow in rough rock fractures PhD Thesis,* School of Civil Engineering, University of Wollongong.

Pullan, A. J., (1990) The quasi-linear approximation for unsaturated porous-media. *Water Resources Research,* 26, 1219{1234, 1990.

Quadros E.F. (1982) *Flow of water in rock joints.* Dissertation. MSc. Degree. Polytechnical School. São Paulo University, 1982. (In Portuguese).

Raven, K.G., Novakowski, K.S. and Lapcevic, P.A. (1988) Interpretation of field tests of a single fracture using a transient solute storage model. *Water Resour. Res.* 24 (12): 2019-2032.

Renard Ph. & de Marsily G. (1997) Calculating equivalent permeability: a review Advances in *Water Resources.* Vol. 20, no. 5-6, p. 253-278.

Renshaw C.E. (1998) Sample bias and the scaling of hydraulic conductivity in fractured rock. *Geophys. Resear. Lett.,* 25(1):121–124.

Richards, L.A. (1931) Capillary conduction of liquids through porous mediums. Physics. *1 (5): 318-333.*

Ruland W.W., Cherry J.A. and Feenstra S. (1991) The depth of fractures and active ground-water flow in a clayey till plain in southwestern Ontario. *Ground Water,* 29(3):405–417.

Ruqvist J. & Stephansson O. (2003) The role of hydromechanical coupling in fractured rock engineering. *Hydrogeology Journal.* 11:7–40 (2003).

Sara M. (2003) *Site Assessment and remediation Handbook 2nd Edition.* Published by Lewis Publishers. ISBN 0-56670-577-0.

Sarkar S., M. Nafi Toksöz, Daniel R. Burns (2004) *Fluid Flow Modeling in Fractures.* Earth Resources Laboratory Dept. of Earth, Atmospheric and Planetary Sciences Massachusetts Institute of Technology Cambridge, MA 02139 Internal Report.

Schulze-Makuch, D. & D. Cherkauer, (1995) Facies dependent scale behavior of hydraulic conductivity and longitudinal dispersivity: in *Groundwater Quality: Remediation and Protection, IAHS Publ.* No. 225, p 157 - 164.

Shapiro, A. M., & Hsieh P. A. (1994) Overview of research at the Mirror Lake site: use of hydrologic, geophysical and geochemical methods to characterize flow and transport in fractured rock. In U.S. Geological Survey Toxic Substances Hydrology Program—*Proceedings of the Technical Meeting, Colorado Springs, Colorado,* September 20-24, 1993, D.W. Morganwalp and D.A. eds. USGS Water Resources Investigation Report 94-4015, U.S. Geological Survey, Reston, Va.

Singhal B.B.S. & Gupta R.P. (1999) *Applied Hydrogeology of Fractured Rocks.* Kluwer Academic Publishers Dordrecht, The Netherlands, 400 pp.

Singhal B.B.S. (2008) Nature of Hard Rock Aquifers: hydrogeological Uncertainties and Ambiguities In: *Groundwater Dynamics in Hard Rock Aquifers: Sustainable Management and Optimal Monitoring Network Design* edited by Abdin Salih, R. Jayakumar Springer, 2008 ISBN 1402065396, 9781402065392 251 pp.

Skagius, K. & Neretnieks I. (1986) Porosities and Diffusivities of some Nonsorbing Species in Crystalline Rocks, *Water Resources Research,* 22(3): 389-398.

Skagius K., (1986) *Diffusion of dissolved species in the matrix of some Swedish crystalline rocks.* Doctoral thesis at the Royal Institute of Technology, Stockholm, Sweden.

Snow D.T. (1968) Rock fracture spacings, openings and porosities. Journal of the Soil Mechanics and Foundations Division, *Proceedings of the American Society of Civil Engineers.* 94, 73–91.

Tang, D.H., Frind, E.O. and Sudicky, E.A. (1981) Contaminant transport in fractured porous media: Analytical solution for a single fracture. *Water Resour. Res.* 17 (3): 555-564.

Thiel K. (1989) Rock Mechanics in Hydroengineering. *Developments in Geotechnical Engineering,* Vol. 51. Elsevier, Amsterdam, 408 p.

Tiedeman C.R., Goode D.J. and Hsieh P.A. (1998) Characterizing a ground water basin in a New England mountain and valley terrain. *Ground Water,* 36:611–620.

Tran N. H. (2004) *Characterisation and modelling of naturally fractured reservoirs* PhD dissertation, University of New South Wales, Sydney, Australia.

Tsang Y.W., Tsang C.F. and Hale F.V. (1996) Tracer transport in a stochastic continuum model of fractured media. *Water Resources Research,* 32(10):3077–3092.

Tsang Y.W. & Tsang C.F. (1987) Channel model of flow through fractured media. *Water Resources Research,* **23(3)**, 467-479.

Tsang C.F & Neretnieks, I. (1998) Flow channelling in heterogeneous fractured rocks. *Rev. Geophys.* 36: 275-298.

Tsang, Y.W., Tsang, C.F., Neretnieks, I. and Moreno, L. (1988) Flow and tracer transport in fractured media: a variable aperture channel model and its properties. *Water Resour. Res.* 24:2049-2060.

Tsang C.F. (1993) Tracer Transport in Fractured Systems.- in: Bear, J., Tsang, C.-F. & de Marsily, G. (Eds.) (1993): *Flow and Contaminant Transport in Fractured Rock*: 237 - 266; New York (Academic Press Inc.).

Tsang, Y.W. & Tsang, C.F. (1989) Flow Channelling in a Single Fracture as a twodimensional strongly heterogeneous permeable medium. *Water Resour. Res.* 25 (9): 2077-2080.

U.S. Environmental Protection Agency (EPA) (1996) *Pump-and-Treat Ground-Water Remediation A Guide for Decision Makers and Practitioners* EPA/625/R-95/005.

Van Golf-Racht, T.D. (1982) *Fundamentals of Fractured Reservoir Engineering. Developments in Petroleum Science*, no. 12, Elsevier Scientific Publishing Company, Netherlands.

Wang, J. S. Y. & Narasimhan T. N. (1985) Hydrologic mechanisms governing fluid flow in a partially saturated, fractured, porous medium. *Water Resources Research.* 21(12): 1861-1874. Copyright by American Geophysical Union.

Witherspoon P.A., Wang J.C.Y., Iway K. and Gale J.E. (1980) Validity of the cubic law for fluid flow in a deformable rock fracture. *Water Resources Research.* 1980; 16-6: 1016-1024.

Witthüser, K.T. (2001) *Untersuchung zum Stofftransport in geklüfteten Festgesteinen unter besonderer Berücksichtigung der Matrixdiffusion.* Ph.D Thesis, Department of Applied Geology, University of Karlsruhe, Germany.

Zimmerman R.W. & Bodvarsson G.S. (1996) Hydraulic conductivity of rock fractures. *Transport in Porous Media,* Vol: 23, Pages: 1 - 30.

Zimmerman, R.W. & Main, I.G. (2004) Hydromechanical behaviour of fractured rocks, In: Y Guéguen and M Boutéca, editor, *Mechanics of Fluid-Saturated Rocks,* London and San Diego, Academic Press, 2004, Pages: 363 - 421, ISBN: 0-12-305355-2.

Zimmerman R.W. & Yeo I.W. (2000) Fluid flow in rock fractures: From the Navier-Stokes equations to the cubic law, In: B Faybishenko, S Benson, and P Witherspoon, editor, *Dynamics of Fluids in Fractured Rocks*, Washington, DC, American Geophysical Union, Pages: 213 - 224, ISBN: 0-87590-9809.

INDEX

A

B

D

E

F

M

N

O

P

Q

R

S

T